T0271265

Wireless Discrete-Time Receivers

This book is the first comprehensive guide to discrete-time (DT) receivers (RX), discussing the fundamental concepts and implications of the technology. It will serve as an essential reference, covering the necessary building blocks of this field, such as low-noise transconductance amplifiers, current-driven mixers, DT band-pass filters, and DT low-pass filters. As well as addressing the basics, the authors present the most recent state-of-the-art techniques applied to the DT RX blocks. A step-by-step style is used to allow readers to develop the required skills to design the DT receivers at the architectural level while providing in-depth knowledge of the details. Written by leading experts in the fields of academia, research, and industry, this book provides an excellent reference to the subject for a wide audience, from postgraduate students to experienced researchers and professionals working with RF circuits.

Massoud Tohidian is currently Chief Technology Officer and a managing partner at Qualinx BV, a high-tech fabless semiconductor company using digital RF techniques. He has been involved in developing low-power CMOS wireless chips for the company. Dr. Tohidian has a PhD from Delft University of Technology in electrical engineering.

Iman Madadi is CEO and cofounder of Qualinx BV, a high-tech fabless semiconductor company using digital RF techniques. Dr. Madadi holds a PhD from Delft University of Technology in electrical engineering.

Amir Bozorg is currently a postdoctoral research scientist at University College Dublin and at Equal1 Labs. Dr. Bozorg holds a PhD from University College Dublin in electrical engineering and has authored several IEEE journal papers and holds four issued US patents in the field of RF-CMOS design.

Robert Bogdan Staszewski is a full professor of electronic circuits at University College Dublin and a guest professor at Delft University of Technology. Professor Staszewski is an IEEE Fellow and a recipient of the 2012 IEEE Circuits and Systems Industrial Pioneer Award. He is also a cofounder of Equal1 Labs. He spent 14 years at Texas Instruments designing wireless transceivers based on the presented technology.

Wireless Discrete-Time Receivers

MASSOUD TOHIDIAN
Qualinx BV, Delft

IMAN MADADI
Qualinx BV, Delft

AMIR BOZORG
University College Dublin

ROBERT BOGDAN STASZEWSKI
University College Dublin

CAMBRIDGE
UNIVERSITY PRESS

University Printing House, Cambridge CB2 8BS, United Kingdom

One Liberty Plaza, 20th Floor, New York, NY 10006, USA

477 Williamstown Road, Port Melbourne, VIC 3207, Australia

314–321, 3rd Floor, Plot 3, Splendor Forum, Jasola District Centre, New Delhi – 110025, India

103 Penang Road, #05–06/07, Visioncrest Commercial, Singapore 238467

Cambridge University Press is part of the University of Cambridge.

It furthers the University's mission by disseminating knowledge in the pursuit of
education, learning, and research at the highest international levels of excellence.

www.cambridge.org
Information on this title: www.cambridge.org/9781107194700
DOI: 10.1017/9781108163620

First published 2022

A catalogue record for this publication is available from the British Library.

Library of Congress Cataloging-in-Publication Data
Names: Tohidian, Massoud, 1985– author.
Title: Wireless discrete-time receivers : monograph / Massoud Tohidian, Qualinx B.V., Delft,
    CTO and Co-founder, Iman Madadi, Qualinx B.V., Delft, CTO and Co-founder,
    Amir Bozorg, University College Dublin, School of Electrical and Electronic Engineering,
    Robert Bogdan Staszewski, University College Dublin, School of Electrical and
    Electronic Engineering.
Other titles: Discrete-time receivers
Description: New York, NY : Cambridge University Press, 2022. |
    Includes bibliographical references and index.
Identifiers: LCCN 2021056116 (print) | LCCN 2021056117 (ebook) |
    ISBN 9781107194700 (hardback) | ISBN 9781108163620 (epub)
Subjects: LCSH: Radio–Transmitter-receivers–Design and construction. |
    Wireless communication systems–Equipment and supplies–Design and construction. |
    Timing circuits–Design and construction. | Discrete-time systems.
Classification: LCC TK6564.3 .T64 2022 (print) | LCC TK6564.3 (ebook) |
    DDC 621.384–dc23/eng/20220113
LC record available at https://lccn.loc.gov/2021056116
LC ebook record available at https://lccn.loc.gov/2021056117

ISBN 978-1-107-19470-0 Hardback

To my parents

— Amir Bozorg

To my parents: Kazimierz and Irena.
To Sunisa, Alexander, and Erik.

— Robert Bogdan Staszewski

Contents

Preface

The ideas described in this book go back to the year 2000, when, together with my Texas Instruments (TI) colleague and research collaborator, Dr. Khurram Muhammad, and inspired by our luminary physicist colleague, Dr. Dirk Leipold, with the tacit approval of our managers, Ken Maggio and Dr. Bill Krenik (TI's motto back then was "Do not ask for permission, ask for forgiveness"), we conceived an idea of realizing the RF reception functionality using discrete-time (DT) charge packets that can be processed using mostly passive switched-capacitor circuitry. This was motivated by the fact that TI was then the unquestionable leader in CMOS process development for low-leakage (i.e., mobile) digital processors, but with an almost nonexistent market presence in RF transceivers. To leapfrog the formidable competitors, such as Infineon and ST Microelectronics, and to secure continued business with Nokia, we had to come up with a distinctly different approach from the conventional continuous-time analog-intensive implementations, something that would exploit TI's leadership in digital CMOS technology. Our invention was the DT receiver (RX) front end that was amenable to the purely digital CMOS process that would not require any analog extensions. The DT RX was quickly put into high-volume production for Bluetooth and GSM single-chip radios. Later on, we had discovered to our surprise that Professor. Abidi's group at the University of California–Los Angeles was working in parallel on a similar research.

This new field was generating lots of exciting ideas. Not unexpectedly, the business product environment is known to not always be terribly welcoming of new ideas, especially when the original DT RX ideas were just "good enough" to secure new products and markets. In that environment, I left TI and moved to Europe to accept a fully tenured faculty position at Delft University of Technology in the Netherlands, where I joined Professor John Long's team. While there, I was finally free to realize the many ideas that were just begging to be explored. After 20 years in industry, I soon realized that the grass on the other (i.e., academia) side was actually not that much greener. All ideas would require funding to pay for the high cost of research. I was lucky to quickly find a sponsor: Mr. Zhuobiao He from the HiSilicon group of Huawei in Shanghai. This allowed me to hire two young, bright-eyed PhD students, Massoud Tohidian and Iman Madadi, who did a fantastic job of realizing the world's first discrete-time superheterodyne radio for wireless cellular applications. The effort was further continued in collaboration with Dr. Patrick Vandenameele from the M4S group of Huawei in Leuven, Belgium, to build commercial 4G/5G radios.

While Tohidian and Madadi were busy writing their PhD dissertations and getting ready to start their own company, Qualinx b.v. where they tried to produce receiver for satellite navigation, a visiting PhD student from Brazil, Dr. Sandro Binsfeld-Ferreira, took it upon himself to realize these DT superheterodyne RX ideas in ultra-low-power applications. Together with our research collaborator from TSMC in Hsin-Chu, Taiwan, Dr. Feng-Wei Kuo, who later became my part-time PhD student, they developed the most power-efficient Bluetooth low-energy receiver for Internet-of-Things (IoT) applications.

In 2014, I moved to University College Dublin in Ireland, where I hired another bright-eyed PhD student, Amir Bozorg, to continue developing ideas along the lines of DT RX. Some of the developed circuits and architectures are included in this book, but I believe the best is yet to come. So, stay tuned . . . .

This book consists of five chapters. Chapter 1 describes the fundamentals of the DT RF receiver architectures and how the sampling phenomenon affects their performance. Moreover, it explains the key concepts of DT filtering.

Chapter 2 deals with the low-noise transconductance amplifier (LNTA), giving an overview on wideband LNTA and noise cancelation techniques, noise and gain analysis.

Chapters 3 and 4 address high-order, low-pass infinite-impulse response (IIR) and band-pass complex charge-sharing filter architectures, respectively. The main goal is to show that high-order low-noise filters can be developed by applying the switch-capacitor circuitry.

Finally, Chapter 5 takes the reader through case studies of different types of DT receivers that exemplify the DT approach taken in the actual RF receivers.

I would also like to thank the staff at Cambridge University Press, particularly Sara Strange and Julia Ford, for their support and patience.

—*Robert Bogdan Staszewski (on behalf of co-authors)*

1 Fundamentals of Discrete-Time RF Receivers

We start the book with the basics. In this chapter, we first present the motivation and fundamentals of discrete-time (DT) radio-frequency (RF) signal processing, and an overview of zero/low intermediate-frequency (IF) and superheterodyne receiver architectures. Then, different sampling schemes present in the state-of-the-art zero-IF DT receivers are studied using a simplified DT receiver. At the end, a $4\times$-sampling concept is introduced for use in DT high-IF receivers [1].

1.1 Why Discrete Time?

While the main motivations of CMOS scaling have been to reduce transistor cost and to improve digital performance, conventional RF/analog designs have not benefited significantly. A finer process node produces shorter digital gate delays while a lowered supply voltage and gate capacitance reduce power consumption. As shown in Fig. 1.1, from 180 to 28 nm CMOS, the V_{DD} supply is reduced more than 30% while the MOS threshold voltage (V_{th}) has not changed considerably. Therefore, the precious available voltage headroom for RF/analog design is now reduced dramatically [2]. Considering also the reduced MOS intrinsic gain [2] and its saturation linearity in scaled CMOS [3], continuous-time (CT) RF/analog design is becoming generally more difficult. In this way, the power consumption and area of the traditional RF/analog designs are not directly process scalable.

On the other hand, the majority of cellular and wireless standard frequency bands are allocated from 400 MHz to 6 GHz, and have not significantly changed for many years. Meanwhile, the transistor cutoff frequency (f_T) has improved dramatically with scaling, as shown in Fig. 1.1. For example, the period from 1999 to 2011 has seen f_T increasing from 20 GHz in 0.35 μm to more than 400 GHz in 28 nm process. This suggests that conventional CT techniques that were optimized for the older technology do not effectively use the ultrahigh speed of transistors of scaled CMOS to improve the performance of RF/analog designs.

In contrast, the newly introduced discrete-time (DT) RF/analog blocks (Fig. 1.2) avoid using complicated traditional analog components such as opamps, and most of signal processing and filtering are done using passive switched-capacitor circuits [4, 5]. Waveforms required for driving the switches are also generated using digital logic. To provide signal gain, DT techniques use inverter-based g_m-cells that avoid

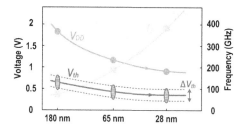

Figure 1.1 Typical CMOS scaling trends for low-power/low-leakage process technology.

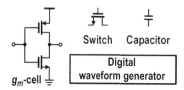

Figure 1.2 Components used in DT signal processing.

transistor stacking and are always compatible with digital technology. As the technology scales, MOS switches become faster and tinier with lower parasitic capacitances. Digital waveform generator becomes also faster and more power efficient. Moreover, the metal capacitor density improves from one process node to the next, resulting in a reduced area. In addition, the inverter-based g_m-cell structure is fully scalable with improved g_m over its bias current. In this way, DT receivers directly benefit from scaling similar to that in digital circuits. References [6–15] are examples of DT process-scalable receivers.

1.2   Overview of Wireless Receiver Architectures

The pioneers of RFIC integration [16] have quickly realized the superiority of operating receivers at zero-IF (ZIF) and low-IF (LIF) rather than at high-IF (HIF): simpler architecture, and a much higher level of monolithic integration as a result of using low-frequency low-pass filters (LPF) for channel selection (see Fig. 1.3(a)). This was despite the many issues associated with ZIF/LIF receivers: time-variant dc offsets, sensitivity to $1/f$ (flicker) noise, large in-band local oscillator (LO) leakage and the second-order nonlinearity [17–22]. The LO leakage to the low-noise amplifier (LNA) input is amplified and then mixed with the LO again, creating a dc offset. This offset could be up to two to three orders of magnitude larger than the wanted signal at the mixer output [23]. By considering the LO leakage to antenna, it could be radiated and subsequently reflected from a moving subject back to the antenna. In this case, the dc offset is time varying and thus much harder to be canceled. In general, all ZIF/LIF issues were viewed rather as an inconvenience and handled through various calibrations. However, high-performance cellular ZIF/LIF receivers now require extensive

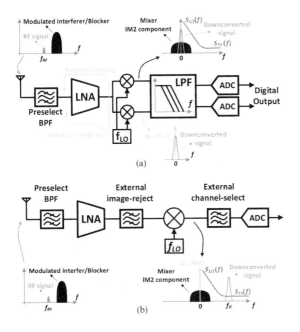

Figure 1.3 Comparison of conventional receiver architectures: (a) zero-IF/low-IF and (b) superheterodyne.

calibration efforts. For example: an intensive IIP2 calibration needs to be concurrently run in the background with dc offset and harmonic rejection (HR) calibration [24, 25].

A superheterodyne architecture, shown in Fig. 1.3(b), pushes the IF frequency much higher such that the aforementioned problems are not a major concern anymore. Despite the obvious advantages, the superheterodyne radios were abandoned decades ago because it was extremely difficult to integrate a high quality (Q)-factor BPF for image rejection and channel selection in CMOS using continuous-time (CT) circuitry [16].

Furthermore, conventional multiband, multistandard cellular receivers (RXs) require many external duplexers, surface acoustic wave (SAW) filters, and switches, typically one per band, to attenuate out-of-band (OB) blockers before they reach the sensitive LNA's input. In time-division duplexing (TDD) systems, external SAW filters can be eliminated if the RX chain could handle large interferers (e.g., 0 dBm at 20 MHz away from a GSM channel of interest [26]). On the other hand, for frequency-division duplexing (FDD) systems, the external SAW filters are responsible for not only the filtering of out-of-band blockers but also for duplexing, that is, separation of concurrent transmit (TX) and RX operations. To reduce the cost and size of the total system solution, in which the external antenna interface network is nowadays the largest contributor, the recent trend is to eliminate SAW filters and switches by using a highly linear wideband RX [27, 18–22]. As a consequence, the isolation of TX-to-RX, and the suppression of TX interferers are worsening, which all further increase the RX linearity requirements in FDD systems.

The resulting reduction in out-of-band filtering implies tough IIP2 requirements (e.g., 90 dBm [25, 22]) for ZIF and LIF receivers. The IIP2 performance of such receivers depends mainly on the second-order nonlinearity of LNA and RF mixer in the receiver chain, as shown in Fig. 1.3(a). Since the typical IIP2 of an RF mixer is between 50 and 70 dB [28], ZIF/LIF receivers require highly sophisticated calibration algorithms [29–33, 22, 34] to be frequently executed to account for variations in, V_{DD}, [19, 35, 36, 24, 37, 38], process corner [38], temperature [39], mixer transistor's gate bias [35], RF blocker frequency [33, 36–38], LO frequency [36–38], LO power [38], and channel frequency [39]. Also, the IIP2 calibration time is rather very slow to find optimum setting for the mixer, and it needs to be run repeatedly due to environmental and operational changes [35].

Most of the filtering and amplification in a zero-IF receiver are done after the mixer, at low frequency. In CMOS implementations, the flicker noise of devices at low frequencies corrupts the wanted signal, leading to a higher noise figure (NF) of the receiver. In contrast, filtering and amplification in a superheterodyne are done normally at higher frequencies than the device's flicker corner.

Superheterodyne or HIF architectures, on the other hand, can have a theoretically infinite IIP2. As shown in Fig. 1.3(b), the desired signal and modulated blocker at the RF input will be downconverted to a higher IF and dc, respectively; thus the modulated blocker can be completely filtered out by a band-pass filter (BPF) [40, 41]. For this reason, there is an increasing interest in uncalibrated high-IIP2 SAW-less superheterodyne RXs with integrated blocker-tolerant BPFs that are amenable to CMOS scaling.

1.3   Discrete-Time Concepts

1.3.1   Direct-Sampling Mixer

The basic idea of the current-mode direct-sampling mixer [42, 43] is illustrated in Fig. 1.4(a). The low-noise transconductance amplifier (LNTA) converts the received RF voltage v_{RF} into i_{RF} in current domain through the transconductance gain g_m. The current i_{RF} gets switched by the half-cycle of the local oscillator (LO) and

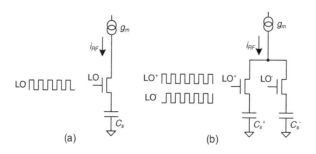

Figure 1.4 Temporal MA operation at RF rate: (a) single-ended and (b) pseudo-differential configurations.

integrated into the sampling capacitor C_s. Since it is difficult to switch the current at RF rate, it could be merely redirected to an identical sampler that is operating on the opposite half-cycle of the LO clock, as shown in Fig. 1.4(b) for a pseudo-differential configuration.

If the LO oscillating at f_0 frequency is synchronous and in phase with the sinusoidal RF waveform, the voltage gain of a single RF half-cycle is

$$G_{v,RF} = \frac{1}{\pi} \cdot \frac{1}{f_0} \cdot \frac{g_m}{C_s}, \tag{1.1}$$

and the accumulated charge on the sampling capacitor is

$$G_{q,RF} = \frac{1}{\pi} \cdot \frac{1}{f_0} \cdot g_m. \tag{1.2}$$

In (1.1) and (1.2), the $\frac{1}{\pi}$ factor is contributed by the half-cycle sinusoidal integration. As an example, if $g_m = 30$ mS, $C_s = 15.925$ pF, and $f_0 = 2.4$ GHz, then $G_{v,RF} = 0.25$.

1.3.2 Temporal Moving Average

Continuously accumulating the charge as shown in Fig. 1.4 is not very practical if it cannot be read out. In addition, a mechanism to prevent the charge overflow is needed. Both of these operations are accomplished by fixing the integration window length followed by a charge readout phase that will also discharge the sampling capacitor such that the next period of integration would start from the same zero condition. The RF sampling and readout operations are cyclically rotated on both C_s capacitors as shown in Fig. 1.5. When LO_A rectifies N RF cycles that are being integrated on the first sampling capacitor, LO_B is off and the second sampling capacitor charge is being read out. On the following N RF cycles, the operation is reversed. This way, the charge integration and readout occur at the same time and no RF cycles are missed.

The sampling capacitor integrates the half-rectified RF current over N cycles. The charge accumulated on the sampling capacitor and the resulting voltage ($V = Q/C_s$) increases with the integration window, thus giving rise to a discrete signal processing gain of N.

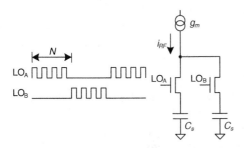

Figure 1.5 Temporal MA operation at RF rate with cyclic charge readout.

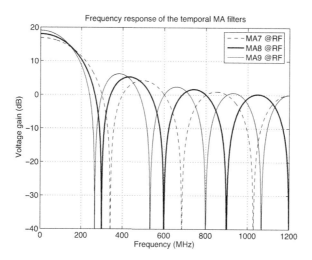

Figure 1.6 Transfer function of the temporal MA operation at RF rate.

The temporal integration of N half-rectified RF samples performs a (FIR) operation with N all-one coefficients, also known as moving average (MA), according to the equation:

$$w_i = \sum_{l=0}^{N-1} u_{i-l},\tag{1.3}$$

where u_i is the ith RF sample of the input charge sample and w_i is the accumulated charge. Since the charge accumulation is done on the same capacitor, this formula could also be used in the voltage domain. Its frequency response is a *sinc* function and is shown in Fig. 1.6 for $N = 8$ (solid line) and $N = 7, 9$ (dotted lines) with sampling rate $f_0 = 2.4$ GHz. It should be noted that this filtering is done on the same capacitor in time domain, resulting in a most *faithful* reproduction of the transfer function.

Due to the fact that the MA output is being read out at the lower rate of N RF clock cycles, there is an additional aliasing with foldover frequency at $f_0/2N$ and located halfway to the first notch. Consequently, the frequency response of MA $= 7$ with decimation of 7 exhibits less aliasing and features wider notches than MA $= 8$ or MA $= 9$ with decimation of 8 or 9, respectively.

It should be emphasized that the voltage G_v and charge G_q signal processing gains of the temporal moving average (TMA) (followed by decimation) are merely due to the sampling time interval expansion of this discrete-time system (the sampling rate of the input is at the RF frequency): $G_{v,tma} = G_{q,tma} = N$.

In the following analysis, the RF half-cycle integration voltage gain of $\frac{g_m}{\pi C_s f_0}$ is tracked separately. Since this gain depends on the absolute physical parameters of normally low tolerance (g_m value of the preceding LNTA stage and the total integrating capacitance of the sampling mixer), it is advantageous to keep it decoupled from the discrete signal processing gain of the multi-tap direct-sampling mixer (MTDSM).

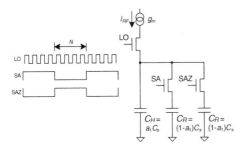

Figure 1.7 IIR operation with cyclic charge readout.

1.3.3 High-Rate IIR Filtering

Figure 1.5 is now modified to include recursive operation that gives rise to the IIR filtering capability, which is generally considered stronger than that of FIR.

A "history" sampling capacitor C_H is added in Fig. 1.7. The integration operation is continually performed on the history capacitor $C_H = a_1 C_s$ and one of the two rotating "charge-and-readout" capacitors $C_R = (1 - a_1)C_s$ such that the total RF integrating capacitance, as seen by the LNTA, is always $C_H + C_R = C_s$. When one of the C_R capacitors is being used for readout, the other is being used for RF integration.

The IIR filtering capability comes into play in the following way: The RF current is being integrated over N RF cycles, as described before. This time, the charge is being shared on both C_H and C_R capacitors proportionately to their capacitance values. At the end of the accumulation cycle, the active C_R capacitor, that stores $(1 - a_1)$ of the total charge, stops further accumulating in preparation for charge readout. The other rotating capacitor joins the C_H capacitor in the RF sampling process and, at the same time, obtains $\frac{1-a_1}{a_1+(1-a_1)} = 1 - a_1$ of the total remaining charge in the "history" capacitor, provided it has no initial charge at the time of commutation. Thus, the system retains a_1 portion of the total system charge of the previous cycle.

If the input charge accumulated over the most-recent N RF samples is w_j then the charge s_j stored in the system at sampling time j, where $i = N \cdot j$, (as stated earlier, i is the RF cycle index) could be described as a single-pole recursive IIR equation:

$$s_j = a_1 s_{j-1} + w_j, \tag{1.4}$$

$$x_j = (1 - a_1)s_{j-1}, \tag{1.5}$$

$$a_1 = \frac{C_H}{C_H + C_R}. \tag{1.6}$$

The output charge x_j is $(1 - a_1)$ of the system charge in the most recent cycle. This discrete-time IIR filter operates at f_0/N sampling rate and introduces a single pole with the frequency attenuation of 20 dB/dec. The equivalent pole location in the continuous-time domain for $f_{c1} \ll f_0/N$ is

$$f_{c1} = \frac{1}{2\pi} \frac{f_0}{N} \cdot (1 - a_1) = \frac{1}{2\pi} \frac{f_0}{N} \cdot \frac{C_R}{C_H + C_R}. \tag{1.7}$$

Since there is no sampling time expansion for the IIR operation, the discrete signal processing charge gain is one. In other words, due to the charge conservation principle, the input charge per sample interval is on average the same as the output charge. For the voltage gain, however, there is an impedance transformation of $C_{input} = C_s$ and $C_{output} = (1 - a_1)C_s$, thus resulting in a gain.

$$G_{q,iir1} = 1, \tag{1.8}$$

$$G_{v,iir1} = \frac{1}{1 - a_1} = \frac{C_H + C_R}{C_R}. \tag{1.9}$$

As an example, the IIR filtering with a single coefficient of $a_1 = 0.9686$, placing the pole at $f_{c1} = 1.5$ MHz ($C_R = 0.5$ pF, $C_H = 15.425$ pF) is performed at $f_0/N = 2.4$ GHz / 8 = 300 MHz sampling rate, and it follows the FIR MA = 8 filtering of the input at f_0 RF sampling rate. The voltage gain of the high-rate IIR filter is 31.85 (30.06 dB).

1.3.4   Additional Spatial MA Filtering Zeros

For practical reasons, it is difficult to read out the x_j output charge of Fig. 1.7 at $f_0/N = 300$ MHz rate. The output charge readout time is extended $M = 4$ times by adding redundancy of four to each of the two original C_R capacitors as shown in Fig. 1.8. The input charge is cyclically integrated within the group of four C_R capacitors. Adding the redundant capacitors gives rise to an additional antialiasing filtering just before the second decimation of M. This could also be considered as equivalent to adding additional $M - 1$ zeros to the IIR transfer function in (1.4).

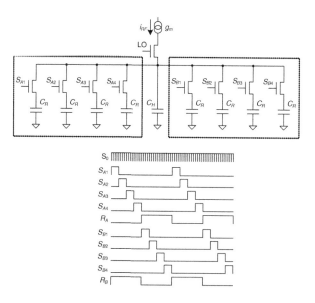

Figure 1.8 IIR operation with additional FIR filtering. The readout and reset circuitry is not shown.

After the first bank of four capacitors gets charged ($S_{A1} - S_{A4}$ in Fig. 1.8), the second bank ($S_{B1} - S_{B4}$) is in the process of being charged and the charge on the first bank of capacitors is summed and read out (R_1). Physically connecting together the four capacitors performs an FIR filtering described as the spatial moving average of $M = 4$:

$$y_k = \sum_{l=0}^{M-1} x_{k-l}, \tag{1.10}$$

where y_k is the output charge and sampling time index $j = M \cdot k$. R_A and R_B in Fig. 1.8 are the readout/reset cycles during which the output charge on the four non-sampling capacitors is transferred out, and the remnant charge is reset before the capacitors are put back into the sampling operation. It should be noted that after the reset phase, but before the sampling phase, the capacitors are unobtrusively precharged [44] in order to implement a dc-offset cancelation or to accomplish a feedback summation for the $\Sigma\Delta$ loop operation.

Since the charge of four capacitors is added, there is a charge gain of $M = 4$ and a voltage gain of 1. Again, as explained before, the charge gain is due to the sampling interval expansion: $G_{q,sma} = M$ and $G_{v,sma} = 1$.

Figure 1.9 shows the frequency response of the temporal moving average with a decimation of 8 ($G_v = 18.06$ dB), the IIR filter operating at RF/8 rate ($G_v = 30.06$ dB), and the spatial moving average filter operating at RF/32 rate ($G_v = 0$ dB) with a decimation of 4. The solid line is the composite transfer function with the dc gain of $G_v = 48.12$ dB. The first decimation of $N = 8$ reveals itself as aliasing. It should be noted that it is possible to avoid aliasing of a very strong interferer into the critical IF band by simply changing the decimation ratio N. This brings out the

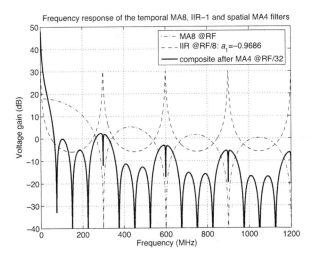

Figure 1.9 Transfer function of the temporal MA filter and the IIR filter operating at RF/8 rate. The solid line is the composite transfer at the output of the spatial MA filter.

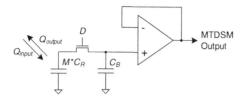

Figure 1.10 Second IIR filter

advantages of integrating RF/analog with digital circuitry by opening new avenues of novel signal processing solutions not possible before.

1.3.5   Lower-Rate IIR Filtering

The voltage stored on the rotating capacitors cannot be readily presented to the MTDSM block output without an active buffer that would isolate the high impedance of the mixer from the required low driving impedance of the output. Figure 1.10 shows the mechanism to realize the second, lower-rate, IIR filtering through passive charge sharing. The active element, the operational amplifier, does not actually take part in the IIR filtering process. It is merely used to sense voltage of the buffer feedback capacitor C_B and present it to the output with a low driving impedance. Figure 1.10 additionally suggests the possibility of differentially combining, through the operational amplifier, the opposite (180° apart) processing path.

The charge y_k accumulated on the $M = 4$ rotating capacitors is being shared during the dumping phase with the buffer feedback capacitor C_B. At the end of the dumping phase, the $M \cdot C_R$ capacitors get disconnected from the second IIR filter and their charge reset before they could be reengaged in the MTDSM operation of Fig. 1.8. This charge loss mechanism gives rise to IIR filtering. If the input charge is y_k, then the charge z_k stored in the buffer capacitor C_B at sampling time k is

$$z_k = a_2(z_{k-1} + y_k) = a_2 z_{k-1} + a_2 y_k. \tag{1.11}$$

$$a_2 = \frac{C_B}{C_B + MC_R}. \tag{1.12}$$

Equation (1.11) describes a single-pole IIR filter with coefficient a_2 and input y_k scaled by a_2, where a_2 corresponds to the storage-to-total capacitance ratio $\frac{C_B}{C_B + MC_R}$. Conversely, due to the linearity property, it could also be thought of as an IIR filter with input y_k and output scaled by a_2.

This discrete-time IIR filter operates at f_0/NM sampling rate and introduces a single pole with the frequency transfer function attenuation of 20 dB/dec. The equivalent pole location in the continuous-time domain for $f_{c2} \ll f_0/(NM)$ is

$$f_{c2} = \frac{1}{2\pi} \frac{f_0}{NM} \cdot (1 - a_2) = \frac{1}{2\pi} \frac{f_0}{NM} \cdot \frac{MC_R}{C_B + MC_R}. \tag{1.13}$$

The actual MTDSM output is the voltage sensed on the buffer feedback capacitor z_k/C_B. The previously used charge stream model cannot be directly applied here because the "output" charge z_k is not the one that leaves the system.

The charge "lost" or reflected back into the $M \cdot C_R$ capacitor for subsequent reset is $(1 - a_2)(z_{k-1} + y_k)$. Due to the charge conservation principle, the time-averaged values of charge input, y_k, and charge leaked out, $(1 - a_2)(z_{k-1} + y_k)$, should be equal. As stated before, the leak-out charge is not the output from the signal processing standpoint. It should be noted that the amplifier does not contribute to the net charge change of the system and, consequently, the only path of the charge loss is through the same $M \cdot C_R$ capacitors being reset after the dumping phase.

The output charge z_k stops at the IIR-2 stage and does not further propagate; therefore, it is of less importance for signal processing analysis. The charge discrete signal processing gain of the second IIR stage is

$$G_{q,iir2} = \frac{a_2}{1 - a_2} = \frac{C_B}{MC_R}. \tag{1.14}$$

The input/output impedance transformation is $\frac{MC_R}{C_B}$. Consequently, the voltage gain of IIR-2 is unity.

$$G_{v,iir2} = 1. \tag{1.15}$$

1.3.6 Cascaded MTDSM Filtering

The cascaded discrete signal processing gain equations of the MTDSM mixer are

$$G_{q,dsp} = G_{q,tma} \cdot G_{q,iir1} \cdot G_{q,sma} \cdot G_{q,iir2} \tag{1.16}$$

$$= N \cdot 1 \cdot M \cdot \frac{C_B}{MC_R} \tag{1.17}$$

$$= \frac{NC_B}{C_R}. \tag{1.18}$$

$$G_{v,dsp} = G_{v,tma} \cdot G_{v,iir1} \cdot G_{v,sma} \cdot G_{v,iir2} \tag{1.19}$$

$$= N \cdot \frac{C_H + C_R}{C_R} \cdot 1 \cdot 1 \tag{1.20}$$

$$= \frac{N(C_H + C_R)}{C_R}. \tag{1.21}$$

Including the RF half-cycle integration (1.1 and 1.2) the total single-ended gain is

$$G_{q,tot} = G_{q,RF} \cdot G_{q,dsp} \tag{1.22}$$

$$= \frac{1}{\pi} \cdot \frac{1}{f_0/N} \cdot g_m \tag{1.23}$$

$$G_{v,tot} = G_{v,RF} \cdot G_{v,dsp} \tag{1.24}$$

$$= \frac{1}{\pi} \cdot \frac{1}{f_0/N} \cdot \frac{g_m}{C_R}. \tag{1.25}$$

Note the similarity between (1.25) and (1.1). In both cases, the term $R_{sc} = \frac{1}{f_s C_s}$ is an equivalent resistance of a switched-capacitor C_s sampling at rate f_s. For example, if $f_s = 300$ MHz and $C_R = 0.5$ pF, then the equivalent resistance is $R_{sc} = 6.7$ kΩ. Since the MTDSM output is differential, the gain values in (1.23)–(1.25) are actually doubled.

The dc-frequency gain $G_{v,tot}$ in (1.25) requires further elaboration. The gain depends only on the g_m of the LNTA stage, rotating capacitor value, and the rotation frequency. Amazingly, it does not depend on the other capacitor values, which contribute only to the filtering transfer function at higher frequencies.

1.3.7   Near-Frequency Interferer Attenuation

Most of the lower-frequency filtering could be realistically done only with the first and second IIR filters. The two FIR filters do not have appreciable filtering capability at low frequencies and are mainly used for antialiasing.

It should be noted that the best filtering could be accomplished by making 3 dB corner frequency of both IIR filters the same and placing them as close to the higher end of the signal band as possible.

$$f_{c1} = f_{c2}. \tag{1.26}$$

This gives the following constraint:

$$C_B = C_H - (M - 1)C_R. \tag{1.27}$$

1.3.8   Signal Processing Example

Figure 1.11 shows the block diagram from the signal processing standpoint for our specific implementation of $f_0 = 2.4$ GHz, $N = 8$, $M = 4$. The following equations describe the time-domain signal processing: (1.3) for w_j, (1.4) and (1.5) for x_j, (1.10) for y_k, and (1.11) for z_k.

The first aliasing frequency (at $f_0/N = 300$ MHz) is partially protected by the first notch of the temporal MA $= 8$ filter. However, for higher-order aliasing and overall

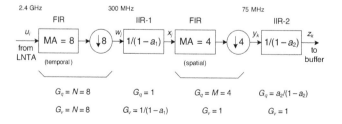

Figure 1.11 Discrete signal processing in the multi-tap direct-sampling mixer (MTDSM).

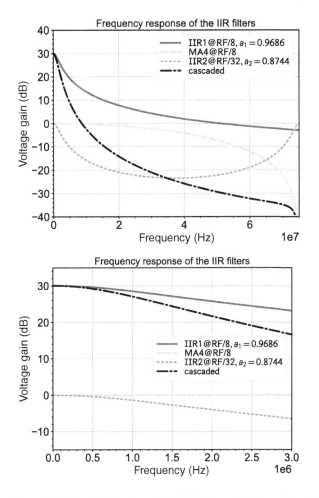

Figure 1.12 Transfer function of the IIR filters with two poles at 1.5 MHz (bottom zoomed).

system robustness, it has to be protected with a truly continuous-time filter, such as an antenna filter. A typical low-cost Bluetooth-band duplexer can attenuate up to 40 dB at 300 MHz offset.

For the above system with an aggressive cut-off frequency of $f_{c1} = f_{c2} = 1.5$ MHz, using $C_R = 0.5$ pF will result in a dc-frequency voltage gain of 63.66 or 36 dB (1.25), and the required capacitance is $C_H = 15.425$ pF (1.7) and $C_B = 13.925$ pF (1.13). The z-domain coefficients of the IIR filters are $a_1 = 0.9686$ and $a_2 = 0.8744$. The dc-frequency gains are $G_{v,iir1} = 31.85$ and $G_{v,iir2} = 1$. The transfer function of these IIR filters is shown in Fig. 1.12. The spatial MA $= 4$, which follows IIR-1, does not appreciably contribute to filtering at lower frequencies but serves as an antialiasing filter for the lower-rate IIR-2. Since the 3 dB point of IIR-2 is slightly corrupted by the discrete-time approximation, the composite attenuation at the cut-off frequencies $f_{c1} = f_{c2} = 1.5$ MHz is about 5.5 dB. The attenuation drops to 13 dB at 3 MHz.

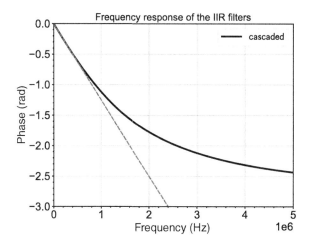

Figure 1.13 Phase response the IIR filters with two poles at 1.5 MHz.

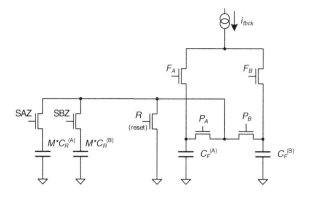

Figure 1.14 Feedback into the rotating capacitors.

Within the 1 MHz band of interest, there is a 3 dB signal attenuation. For the most optimal detector operation, this in-band filtering should be taken into consideration in the matched-filter design. Figure 1.13 shows the phase response of the above structure versus the ideal constant group delay.

1.3.9   MTDSM Feedback Path

The MTDSM feedback correction could be unobtrusively injected into either group of the four rotating capacitors of Fig. 1.8 when they are not in the active sampling state. This way, the main signal path is not perturbed. The feedback correction is accomplished through charge injection/equalization between the "feedback capacitor" C_F and the rotating capacitors C_R in the MTDSM structure by shorting all of them together after the C_R group of capacitors gets reset, but before they are put back to the sampling system. The feedback charge accumulation structure is shown in Fig. 1.14.

Each feedback capacitor C_F is associated with one of the two rotating capacitor of group "A" and "B." The two groups commutate the charging process.

Voltage on the feedback capacitor can be calculated as follows. Charging the feedback capacitor C_F with the current i_{fbck} for the duration of T will result in incremental accumulation of $\Delta Q_{in} = i_{fbck} \cdot T$ charge. This charge gets added to the total charge $Q_F(k)$ of the feedback capacitor at the kth time instance.

$$Q_F(k) = Q_F(k-1) + \Delta Q_{in} = Q_F(k-1) + i_{fbck} \cdot T. \tag{1.28}$$

During the charge distribution moment, the feedback capacitor gets connected with the previously reset group of rotating capacitors $M \cdot C_R$. The charge depleted from C_F is dependent on the relative capacitor values.

$$\Delta Q_{out}(k) = \frac{MC_R}{C_F + MC_R} Q_F(k). \tag{1.29}$$

The charge transferred to the rotating capacitors is proportional to the total accumulated charge Q_F or voltage on the feedback capacitor $V_F = Q_F/C_F$. At first, the accumulated charge is small, so the outgoing charge is small. Since the incoming charge is constant, the Q_F charge will continue accumulating until the net charge intake becomes zero. Equilibrium is reached when $\Delta Q_{in}(k) = \Delta Q_{out}(k)$.

$$i_{fbck} \cdot T = \frac{MC_R}{C_F + MC_R} Q_F(k). \tag{1.30}$$

Transformation of (1.28), (1.29), and (1.30) gives the equilibrium voltage.

$$V_{F,eq} = i_{fbck} \cdot T \cdot \frac{C_F + MC_R}{C_F \cdot MC_R}. \tag{1.31}$$

The $\Delta Q_{out,eq}$ charge transfer into the rotating capacitors at equilibrium will create voltage on the bank of rotating capacitors.

$$V_R = \frac{i_{fbck} \cdot T}{MC_R}. \tag{1.32}$$

As shown in Section 1.3.5, the voltage transfer function from the rotating capacitors to the history capacitor is unity. Therefore, the bias voltage developed on C_H is

$$V_H = \frac{i_{fbck} \cdot T}{MC_R}. \tag{1.33}$$

1.4 DT ZIF Receiver: 1× and 2× Sampling

A simplified conceptual diagram of a DT ZIF receiver is shown in Fig. 1.15(a). The receiver consists of a low-noise transconductance amplifier (LNTA), a pair of quadrature mixers, and two DT sampling low-pass filters. After the antenna, the received RF signal is amplified and converted into the current, i_{RF}, by the LNTA with high output impedance. This current is then downconverted to zero-IF by the quadrature mixers. The mixers are driven by the $LO_{I,Q}$ signals, which are differential 25%

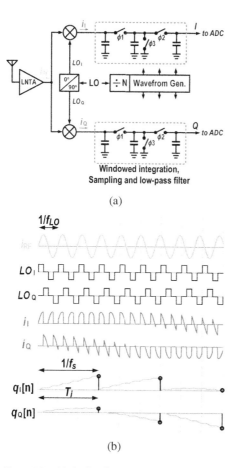

Figure 1.15 (a) A simple DT receiver with passive LPF; and (b) its waveforms at various nodes.

duty-cycle clocks with a 90 phase shift. Considering a narrow-band modulated RF signal, Fig. 1.15(b) shows the signal waveforms at various stages. The current leaving the mixers is integrated over a time window T_i and sampled in the form of DT charge packets [45], $q_{I,Q}[n]$. This DT data is then low-pass filtered by a passive switch-cap circuit (e.g., a second-order IIR [45, 5]). The windowed integration forms a continuous-time (CT) sinc antialiasing filter just before the sampling (Fig. 1.16), and attenuates unwanted signals folded from multiples of the sample frequency f_s (i.e., sampling images) [7, 11]. The window time (and sampling rate) is set straightforwardly by the clock rate of the waveform generator circuit.

In most of the DT ZIF receivers, this sampling is done at a significantly lower rate than the LO frequency (f_{LO}, which is the mixer downconversion frequency) [9–11, 46, 47]. For example, in [11], a sample rate of 480 MS/s is used for a 2.4 GHz RF signal. A lower sample rate increases the attenuation of sampling image frequency at a fixed offset, but creates also extra sampling images at lower offsets [11].

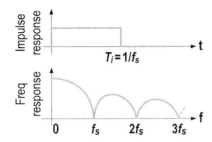

Figure 1.16 Impulse and frequency response of windowed integration.

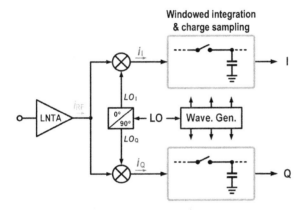

Figure 1.17 Signal sampling in a DT receiver.

1.4.1   1× Sampling in Zero-IF

For the time being, the signal sampling in a DT receiver with the simplified structure of Fig. 1.17 is our focus. Consider the case of a ZIF reception where the signal is sampled at the same rate as the LO frequency [10], hereafter called 1× sampling. Drawn in Fig. 1.18(a), the narrow-band modulated current signal i_{RF} is downconverted as i_I and i_Q baseband quadrature currents, windowed integrated (WI), and then sampled in the form of charge packets, q_{In} and q_{Qn}, at the end of each LO cycle.

A ZIF architecture with 1× sampling has image frequencies at multiples of LO frequency. Figure 1.18(b) shows the frequency translation. The wanted RF signal is downconverted to dc by mixing with the quadrature LO (black tone). At the same time, the frequency bands near zero and $2f_{LO}$ are translated to $\pm f_{LO}$. The windowed integration of i_{IQ} and sampling forms a continuous-time (CT) antialiasing filter (shown in green), with its notches coinciding with the sampling images. The narrower the required bandwidth, the stronger the image attenuation [11]. After the sampling, the attenuated images at multiples of $\pm f_s$ are folded over the wanted signal at baseband.

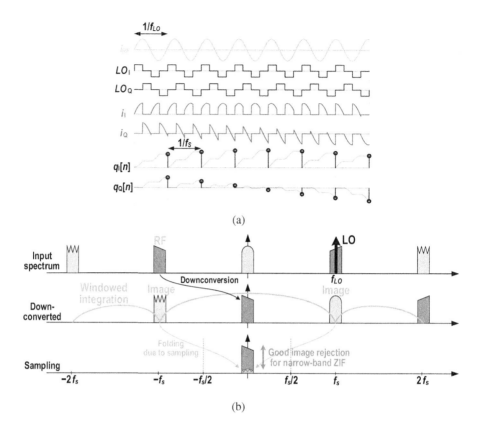

Figure 1.18 (a) Time-domain signal waveforms and (b) frequency translation in a $1\times$ sampling zero-IF DT receiver: input spectrum is shifted to right (RF downconversion) and after windowed integration is sampled.

1.4.2 $2\times$ Sampling in Zero-IF

By increasing the sampling rate to two times the LO frequency (hence, $2\times$ sampling), the ZIF receiver does not introduce any sampling images other than those caused by the mixer's odd harmonics, (i.e., third, fifth, etc.). This should promote wideband reception, which benefits less from the protective notches of the antialiasing WI filter. Figure 1.19(a) shows the transient signal waveforms. The sample rate (f_s) is now doubled with respect to the $1\times$ sampling by reading the integrated currents at twice the rate. It needs to be emphasized that the doubling of sampling rate is independent of the clocks used for the mixer, which is still at f_{LO}.

As shown in Fig. 1.19(b), the antialiasing filter caused by WI is widened twofold. After the sampling at $2\times$, the "Image" bands still remain at high frequency and are not mixed with the wanted signal. Therefore, in the $2\times$ sampling, it is possible to further filter the images prior to decimation and folding over the wanted signal. The only images created by sampling are the self-image of the wanted RF signal and the images that come from odd harmonics of f_{RF} (e.g., $3 f_{RF}$, not shown in the figure),

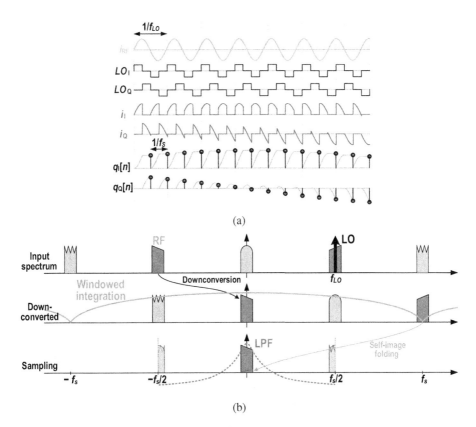

Figure 1.19 (a) Time-domain signal waveforms; and (b) frequency translation in a 2× sampling zero-IF DT receiver. "Image" bands after the sampling are folded on themselves, but remaining apart from the wanted signal and can be filtered afterwards by a DT LPF.

all attenuated earlier by the antialiasing filter. Note that in Fig. 1.19(b), the mixer harmonics that produce the mixer images are not shown for the sake of clarity in illustrating the sampling process.

1.5 4× Sampling for DT High-IF Receiver

1.5.1 2× Sampling in Superheterodyne

If the 2× sampling concept were to be used in a DT superheterodyne receiver, in which the IF frequency (f_{IF}) is high, where $f_{LO} = f_{RF} + f_{IF}$, the receiver would show a poor image rejection. To illustrate that, let us assume spectrum of the input signal as depicted in Fig. 1.20(b). The wanted signal is downconverted to $+f_{IF}$ after the mixer,[1] while part of the image power is upconverted to $2f_{LO} + f_{IF}$. By

---

[1] By assuming I/Q signals are at baseband, the complex signal defined as $I + jQ$ could have an asymmetric spectrum around zero. Moreover, the sampling Nyquist (alias-free) range is from $-f_S/2$ to $f_S/2$ instead of 0 to $f_S/2$ for real signals.

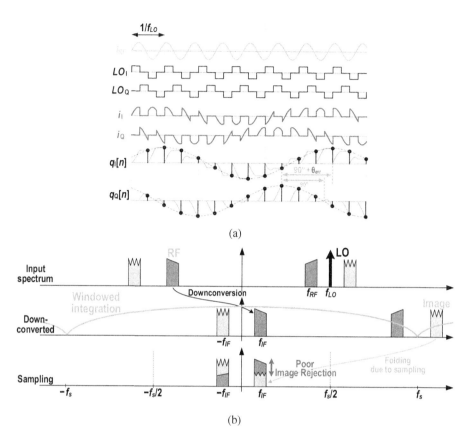

Figure 1.20 (a) Time-domain signal waveforms and (b) frequency translation in a $2\times$ sampling DT superheterodyne receiver. After the sampling, image is aliased on the wanted signal without enough attenuation.

sampling this signal at the $2\times$ rate, this image (whose absolute frequency is higher than the Nyquist rate, $f_s/2$) folds over the wanted signal at $+f_{IF}$. In addition, note that in the superheterodyne the notch of the windowed integration is not aligned (unlike in ZIF) with the image (it is separated by f_{IF}). Therefore, the image signal is not effectively filtered out, and this leads to an unacceptable image rejection of the receiver.

To get further insight, let us inspect the time-domain signals of this structure in Fig. 1.20(a). As expected, the RF signal after downconversion and sampling is centered at a high IF and the two sampled signals (q_{In} and q_{Qn}) have quadrature sinusoidal waveforms. Looking more closely, the phase shift between I and Q is not exactly $90°$, as normally expected for quadrature signals. There is an error of half a cycle of sampling that creates $\theta_{err} = (T_s/2) \times 2\pi f_{IF}$ [48]. This limits the image rejection of this structure.

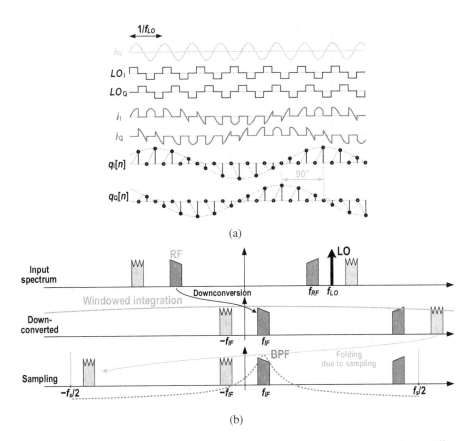

Figure 1.21 (a) Time-domain signal waveforms and (b) frequency translations in a 4× sampling DT superheterodyne receiver. Since f_s is increased to $4f_{LO}$, IF image is completely distinct from the wanted signal and can be filtered afterward by a DT BPF.

1.5.2   4× Sampling Signal Processing Technique

To solve the problem of the DT superheterodyne image introduced by sampling, we proposed advancing to the 4× sampling, that is, $f_s = 4f_{LO}$ [45]. As shown in Fig. 1.21(a), I and Q signals have precisely 90° phase shift after sampling. Although samples with zero value between nonzero samples seem to be noninformative, they are ensuring quadrature accuracy between I and Q.

Considering the signal spectrum in Fig. 1.21(b), the windowed integration filter with its first notch located at $\pm 4f_{LO}$ obviously cannot filter out the image signal. However, this time, the upconverted image (at $2f_{LO} + f_{IF}$) after the sampling folds over $-f_s + f_{IF}$, apart from the wanted signal. Therefore, this signal is not aliased after sampling, thus making a clear frequency separation between the wanted signal and its potential image. Then a DT complex band-pass filter (BPF) is able to select the wanted signal and filter the rest of spectrum (dashed-line transfer function in Fig. 1.21(b)). The only images that are translated directly on top of the wanted signal are the mixer's odd harmonics images. In this respect, further increasing the sample rate (e.g., to $8\times$) without using a harmonic rejection mixer would not be beneficial. In [48], an $8\times$ sampling DT mixing architecture was proposed that also implements harmonic rejection.

To summarize, the $4\times$-sampling concept offers discrete-time signal processing without introducing any unwanted images other than LO odd harmonics. This image-free sampling will later be used in DT receiver examples in Chapter 5.

1.6   Conclusion

The discrete-time manner of RF signal processing is an attractive approach for realizing receivers in an advanced CMOS process technology. It opens up new architectural possibilities, not previously possible with the traditional analog-intensive implementations. Considering the advantages and disadvantages of a superheterodyne architecture compared to zero-IF, and accounting for the recent advancements in the CMOS process technology, it appears that it is now the time to return to the historical superheterodyne for high-performance applications and/or low-power operations. The main remaining challenge is the full CMOS integration of the IF filter, which will be discussed in Chapter 4.

2   First Stage: Low-Noise Transconductance Amplifier

To be able to amplify an RF signal located at any of the supported cellular frequency bands, a wideband noise-canceling low-noise amplifier (LNA) [49] appears to be a good choice. As the receivers, later introduced in Chapter 5, are based on sampling the input charge, the RF amplifier needs to provide *current* rather than voltage, thus acting as a transconductance amplifier (TA) exhibiting a high output impedance compared to the input load of its subsequent stage. An LNTA (i.e., LNA+TA) could trivially be constructed by cascading LNA and TA (g_m) stages [9–12]. However, to improve noise and linearity, both of these circuits should be codesigned and tightly coupled [13, 50]. This chapter presents examples of state-of-the-art wideband noise-canceling LNTAs.

2.1   Introduction

The usage of various wireless standards, such as Bluetooth, Wi-Fi, GPS, and 2G/3G/4G/5G cellular, has been continually increasing. To utilize the frequency bands efficiently and to support more communication standards with lower power consumption, with lower occupied volume, and at reduced costs, multimode transceivers, software-defined radios (SDR), cognitive radios, and so on have been actively investigated [51].

Broadband behavior of a wireless receiver is typically defined by its front-end low-noise amplifier (LNA), whose design must consider trade-offs between input matching, noise figure (NF), gain, bandwidth, linearity, and voltage headroom in a given process technology. There are several wideband LNA design topologies and techniques, including filter-type amplifiers [52], g_m-enhancement technique [53], common-gate (CG) amplifiers [54], resistive shunt-feedback amplifiers [55–57], and distributed amplifiers [58].

A very wide bandwidth LNA can be constructed using a common-source (CS) amplifier topology with several bandpass filters for providing wideband input matching. In [52], a three-section bandpass Chebyshev filter is used to resonate the reactive part of the input impedance to provide wideband input matching over the whole band from 3.1 to 10.6 GHz. However, several of the associated bulky inductors there occupy a large chip area, which makes this technique unsuitable for wideband applications below 3 GHz [58]. Moreover, although the CS configuration typically ensures better noise performance than in a CG structure, a low quality factor (Q) of on-chip

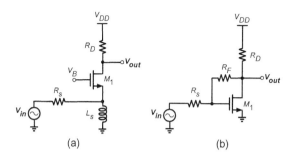

Figure 2.1 Common wideband input-matching techniques: (a) common-gate, and (b) shunt-feedback CS amplifiers.

inductors, especially those at the gate of input stage, deteriorates the noise performance where the minimum achieved NF is limited to 4.2 dB. Distributed amplifiers satisfy the required bandwidth for SDRs and optical communications, but they need several parallel stages to simultaneously provide a sufficiently high bandwidth and gain, thus resulting in high power consumption and large chip area. Moreover, they suffer from high NF due to noise from the gate's line-termination resistors and losses in the inductors [58].

Among the popular techniques for designing wideband LNAs, CG and shunt-feedback CS structures, shown in Fig. 2.1, are of particular interest. The CG stage in Fig. 2.1(a) can realize a broadband input impedance matching without extra components. Since the parasitic gate-drain capacitor there is AC grounded, the CG amplifier has a better input–output isolation than in a shunt-feedback CS amplifier [54]. The linearity of the CG structure is better than that of the CS amplifier, because in the former the input source resistance further provides the source degeneration. The input impedance of the CG structure is roughly $1/(g_{mb1} + g_{m1})$, and the noise factor is $F = 1 + (\gamma/\alpha g_{m1} R_s) + (4/g_{m1} R_D)$ [54], where γ is the excess noise factor in short-channel devices and α is the ratio $\alpha = g_m/g_{ds0}$ of the small-signal transistor transconductance g_m to the zero-bias drain conductance g_{ds0}. g_{mb} models the transistor's body effect. This structure suffers from poor noise performance since its total g_m should be 20 mA/V so as to satisfy the input-matching condition.

A popular method to enhance its noise performance is a noise-cancelation technique provided by a successive stage, which removes the channel thermal noise of the main CG transistor [49]. However, the aggregate noise performance is now limited by the channel thermal noise of the cancelation stage. Finally, another architecture in [59] uses current combining as a means to provide noise cancelation in a receiver that not only cancels the noise due to the antenna input resistance, but also the baseband noise of a transimpedance amplifier (TIA) is up-converted to RF and canceled out there.

In Sections 2.3 and 2.4, two noise-cancelation schemes will be introduced, where in the first structure we further improve upon the aggregate noise performance of the CG architecture with the successive noise-canceling stage by reducing the channel thermal noise of the cancelation stage itself. The key aim is to lower the NF without increasing the consumed power, which is mainly achieved by employing a current-reuse technique. Then, in the second architecture, twofold noise cancelation is introduced, which shows how the noise performance of the CG architecture can be improved while simultaneously providing high gain.

2.2 Overview of Noise-Cancelation and -Reduction Techniques

In this section, we first describe the basic idea of a noise-cancelation scheme. Then, based on that, we introduce a new noise-reduction technique. Finally, these two techniques are combined in a manner that saves power.

2.2.1 Conventional Noise-Cancelation Technique

The most important noise source in CMOS LNAs is the channel thermal noise of MOS transistors. This noise is modeled as a shunt current source across the transistor's drain and source terminals. The designer's goal is to minimize the generation and propagation of this noise. Among various publications introducing noise-cancelation techniques in LNAs, [49, 60–62] are noteworthy.

The conventional noise-cancelation scheme in the CS shunt-feedback topology is shown in Fig. 2.2. The noise current of the main, that is, input-matching, transistor, M_1, flows through the feedback resistor, R_F, toward the M_1 gate and creates two noise voltages at nodes X and Y with the same phase but different amplitudes. On the other hand, the *signal* voltage at these nodes has opposite polarities and different amplitudes due to the inverting operation of the M_1 amplifier. The signal and noise polarities being opposite at nodes X and Y make it possible to cancel the noise originating from the input-matching transistor while adding the signal contributions constructively. The noise voltage at node X, V_{nX}, is amplified and inverted by M_2, while the noise voltage at node Y, V_{nY}, is passed across M_3 barely changed. At the output node, the two voltages with opposite phases are canceled. Ultimately, the channel thermal noise

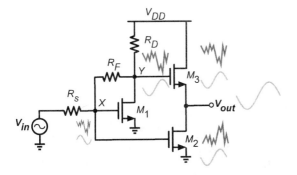

Figure 2.2 Conventional noise cancelation of M_1 configured as a CS shunt-feedback amplifier (biasing not shown).

of M_1 will be greatly attenuated or altogether canceled, provided that the following condition is satisfied:

$$V_{n,out} = V_{nY} \frac{r_{ds2}}{r_{ds2} + 1/g_{m3}} - v_{nX} \frac{g_{m2}}{g_{m3}} = 0$$

$$\frac{R_F + R_s}{R_s} = \frac{g_{m2}}{g_{m3}},$$

(2.1)

where $g_m r_{ds} \gg 1$ was assumed.

As mentioned, this kind of noise cancelation is commonly used in LNA structures with the CS input stage. The main drawback here is the need for an extra stage in order to amplify and invert the voltage noise at node X and add it with the voltage noise at the output. According to (2.1), since the feedback resistor is much larger than the input source resistor, $R_F \gg R_s$, the transconductance of M_2, g_{m2}, must be large enough to satisfy the noise-cancelation condition, but at a cost of higher power consumption. In Section 2.2.2, we offer a new technique that can be used as either a noise-cancelation or noise-reduction technique *without* substantially increasing the power consumption [63].

2.2.2   Noise-Reduction Technique

Technique to Cancel the Noise of Main Transistor

The aforementioned goal of improved noise performance at no extra consumed power can be achieved by means of a current-reuse technique that was inspired by [64, 65]. Figure 2.3 shows this method. Just as in Fig. 2.2, the channel noise of the main transistor M_1 develops a noise voltage at node Y, V_{nY}, which appears on its gate at node X as V_{nX} via the resistive divider attenuation $R_s/(R_s + R_F)$. Likewise, it is then amplified and inverted via M_{aux}. Here, however, the M_{aux}'s current is injected right back into node Y via C_3 to subtract the original noise perturbation in the M_1's

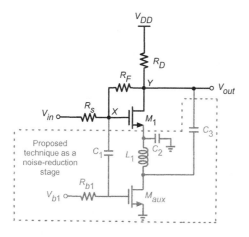

Figure 2.3 The noise-reduction technique of M_1's channel noise.

channel. This way, there is no need for an extra branch M_3 used in the conventional noise cancelation of Fig. 2.2. Furthermore, the source of M_1 is connected to the ground via C_2. Inductor L_1 provides some AC isolation between the source of M_1 and drain of M_2. By stacking M_1 on top of M_2 dc-wise, the dc current is reused, and M_2 is biased by the main transistor current. However, AC-wise, M_{aux} is paralleled with the main transistor M_1 by means of C_1 and C_3, but completes the negative feedback around M_1 for its noise. For the aforementioned technique to cancel the noise of M_1, the following condition should be met:

$$V_{n,out} = V_{nY} - v_{nX} g_{m_{aux}} R_D = 0$$

$$\frac{R_F + R_s}{R_s} = g_{m_{aux}} R_D. \tag{2.2}$$

Equation (2.2) suggests that the full noise-cancellation of M_1 is rather expensive in terms of consumed power, since the ratio of R_F/R_D and $g_{m_{aux}}$ need to be very high.[1] However, this technique could be beneficially used at low expended power for a *partial* noise cancellation, that is, noise *reduction*, of M_1.

Current-Reuse Technique as Noise Reduction

Noise factor excess, F_{M1}, contributed by the M_1 transistor of the shunt-feedback CS amplifier shown in Fig. 2.1(b) is calculated as:

$$F_{M1} = \left| \frac{\overline{V_{n,M1}}/A_v}{\overline{V_{n,Rs}}} \right|^2 = \left| \frac{\overline{I_{n,M1}} Z_{out}}{\overline{V_{n,Rs}} A_v} \right|^2$$

$$= \frac{4kT g_{m1} |Z_{out}|^2}{kT R_s g_{m1}^2 |Z_{out}|^2} \frac{\gamma}{\alpha} = \frac{4}{R_s g_{m1}} \frac{\gamma}{\alpha}, \tag{2.3}$$

where Z_{out} is the output impedance of the amplifier as seen by the unloaded output node. In addition, $\overline{I}_{n,M_1}^2 = 4kT g_{m1} \gamma$ is the channel thermal noise of M_1, and $|A_v| \simeq g_{m1} Z_{out} \cdot Z_{in}/(Z_{in} + R_s)$ is the voltage gain of M_1, where $Z_{in} = R_F/(1 + g_{m1} R_D)||1/sC_{in}$, and C_{in} is due to parasitics at the gate of M_1, for the sake of simplicity Z_{in} is considered equal to R_s. Hence, the noise factor of the shunt-feedback amplifier shown in Fig. 2.1(b) is approximately equal to [56]:

$$F_{(\text{Fig. 2.1b})} \geqslant 1 + \frac{4}{R_s g_{m1}} \frac{\gamma}{\alpha}. \tag{2.4}$$

According to (2.4), the noise factor has a reverse relationship with the transconductance. It means that by increasing the transconductance of the main transistor, the circuit's relative noise contribution is decreased. However, this results in a higher power dissipation.

---

[1] The input matching of the shunt-feedback CS amplifier is defined by R_F and is approximately equal to $R_F/(1 + g_{m1} R_D)$; for providing the noise-cancelation condition, $g_{m_{aux}}$ should be much larger than g_{m1}, that is, $g_{m_{aux}} \simeq (2 + g_{m1} R_D)/R_D$.

By using the discussed current-reuse technique of Fig. 2.3, the noise factor is roughly equal to $F_{(Fig.\ 2.3)} = 1 + F_{M1} + F_{M_{aux}}$, where F_{M1} and $F_{M_{aux}}$ are expressed by:

$$F_{M1} = \frac{4kTg_{m1}|Z_{out}|^2}{kTR_s(g_{m1} + g_{m_{aux}})^2|Z_{out}|^2} \frac{\gamma}{\alpha}$$
$$= \frac{4g_{m1}}{R_s(g_{m1} + g_{m_{aux}})^2} \frac{\gamma}{\alpha}.$$
(2.5)

$$F_{M_{aux}} = \frac{4kTg_{m_{aux}}|Z_{out}|^2}{kTR_s(g_{m1} + g_{m_{aux}})^2|Z_{out}|^2} \frac{\gamma}{\alpha}$$
$$= \frac{4g_{m_{aux}}}{R_s(g_{m1} + g_{m_{aux}})^2} \frac{\gamma}{\alpha}.$$
(2.6)

Finally, the total noise factor of the presented structure, without considering the thermal noise of R_D, is approximately given by:

$$F_{(Fig.\ 2.3)} \geq 1 + \frac{4\gamma}{\alpha(g_{m1} + g_{m_{aux}})R_s} + \frac{4}{R_s R_D(g_{m1} + g_{m_{aux}})^2}.$$
(2.7)

From the standpoint of the received signal, M_{aux} is paralleled with the main transistor M_1, and hence, according to (2.7), their transconductances are summed up. This boost in transconductance reduces the noise figure without increasing the bias current. Without the current-reuse technique, M_{aux} would be paralleled with M_1 in a conventional way as in Fig. 2.2, and the structure would consume twice the power in order to achieve the same NF. Nonetheless, the main drawback of the new technique is the reduced voltage headroom leading to some deterioration of linearity.

To demonstrate the benefit of the noise-reduction technique introduced in Fig. 2.3, we now apply it into the CS noise-canceling LNA of Fig. 2.2 for the purpose of reducing the noise of the latter's second stage (i.e., M_2). To have a better comparison between Figs. 2.2 and 2.5, their respective simplified noise factors, $F_{(Figs.\ 2.2)}$ and $F_{(Fig.\ 2.5)}$, are calculated as follows:

$$F_{(Fig.\ 2.2)} \geq 1 + \frac{\gamma}{\alpha g_{m2}R_s} + \frac{\gamma g_{m3} + \alpha R_D g_{m3}^2}{\alpha R_s g_{m2}^2}.$$
(2.8)

$$F_{(Fig.\ 2.5)} \geq 1 + \frac{\gamma}{\alpha(g_{m2} + g_{m_{aux}})R_s} + \frac{\gamma g_{m3} + \alpha R_D g_{m3}^2}{\alpha R_s(g_{m2} + g_{m_{aux}})^2}.$$
(2.9)

By comparing (2.8) and (2.9), it can be seen that for the same value of g_{m2} and g_{m3} in both structures (Figs. 2.2 and 2.5), the noise performance in Fig. 2.5 has improved.

The efficacy of this noise-reduction technique of Fig. 2.3 is illustrated by the NF circuit simulation plots in Fig. 2.4 with superimposed analytical plots to verify the derived noise equations.[2] It is compared with the basic shunt-feedback amplifier of

---

[2] We extend (2.4) and (2.7) by further considering the thermal noise of R_{D1}, that is,
$F \simeq 1 + [R_D(R_F(1 + (g_{m1} + kg_{m1})R_D))^2/R_s(Z_D + R_F(1 + (g_{m1} + kg_{m1})R_D))^2 Z_D^2(g_{m1} + kg_{m1})^2(R_F + R_F Z_{in}C_{in}s + g_{m1}R_D Z_{in} + Z_{in})^2] + [\gamma/4(R_s(g_{m1} + kg_{m1})(R_s/(R_s(1 + R_s C_{in}s) + R_s)^2)) + (4R_s/R_F)],$

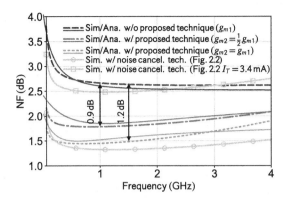

Figure 2.4 Comparison of a simulated/derived NF of the shunt-feedback CS amplifier of Fig. 2.1(b), with the same structure but with the noise-reduction technique of Fig. 2.3, while both structures sink the same current, 1.7 mA, from a 1 V supply. The NF of the conventional noise-cancelation configuration (Fig. 2.2) is included for reference.

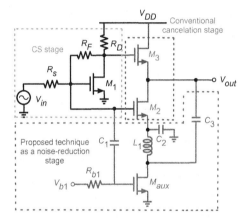

Figure 2.5 Conventional noise-cancelation along with the designed noise-reduction technique.

Fig. 2.1(b) consuming the same power of 1.7 mW. The minimum NF of the basic amplifier is 2.65 dB, while the new technique improves it to 1.45 dB. The obtained NF is now within a small fraction of a dB to the straightforward manner of noise-cancelation shown in Fig. 2.2, but that consumes as much as 10 mW. However, when the current is insufficiently high, not only can the noise of the first stage not be canceled entirely, but it ends up actually adding more noise sources to the circuit, which results in increasing the NF. While we maintain the current of the second stage at 1.7 mA, at the same level as the current of the first stage (the total current of Fig. 2.2 in this case is 3.4 mA), the current of Fig. 2.3 can be just 1.7 mA. As shown in Fig. 2.4, the noise-cancelation technique of this case improves the noise performance slightly

where $k = 0$ gives the result for basic circuit; and also, since R_F is high, its noise effect, $4R_s/R_F$, in the total noise factor is negligible.

(i.e., 0.2 dB). The power efficiency advantages could be summarized as follows: According to (2.1), which describes the conventional noise-cancelation technique, the current of the second stage should be increased in order to satisfy the noise-cancelation condition, resulting in more power drain. Moreover, there are at least two branches in the conventional noise-cancelation technique, which means an extra power consumption because, in addition to the main branch, M_1, the cancelation branch, $M_{2,3}$ drains an extra dc current, while in the noise reduction technique there is only one branch that reuses the dc current for M_1 and M_2.

The salient feature of the noise-*reduction* technique of Fig. 2.3 is that it consists of a single stage, and it saves power by means of the current reuse. This feature allows the structure to be incorporated into the (second) noise-*cancelation* stage of the two-stage amplifier of Fig. 2.2, as illustrated in Fig. 2.5 (another example will be shown in Section 2.3). This way, the channel thermal noise of the noise-canceling device itself (M_2) will be reduced at no extra power. As a net result, the noise-cancelation condition is satisfied more effectively. This is given by:

$$\overline{V}^2_{n,out} = \overline{V}^2_{nY} \left(\frac{r_{ds2}||r_{ds_{aux}}}{r_{ds2}||r_{ds_{aux}} + 1/g_{m3}} \right)^2$$

$$- \overline{V_{nX}}^2 \left(\frac{g_{m2} + g_{m_{aux}}}{g_{m3}} \right)^2 \tag{2.10}$$

$$V_{n,out} = 0 \Rightarrow \frac{R_F + R_s}{R_s} = \frac{g_{m2} + g_{m_{aux}}}{g_{m3}}.$$

In (2.10), g_{m2} is added to g_{m4}, and hence the noise-cancelation condition can be satisfied at lower power. Therefore, applying the noise-reduction approach in the noise-cancelation stage of the conventional noise-cancelation scheme reduces the power dissipation without affecting the NF. Moreover, the added new transistor, M_{aux}, also decreases the noise contribution of the cancelation stage, M_2, without any extra power.

It is worth mentioning that (2.10) is used just to show the beneficial effect of M_{aux} in the conventional noise-cancelation condition, so the parasitic capacitances are not considered. Although, in practice, the condition of (2.10) is not completely satisfied due to the parasitic capacitances and the limitation of power consumption, the noise will be reasonably attenuated even by meeting this condition partially.

2.3   Noise-Reduction Noise-Cancelation LNTA

Section 2.2 introduced the noise-cancelation and -reduction techniques. An example was given in Fig. 2.5 on how they could be beneficially combined to form a noise-canceling LNA in the CS configuration that saves significant power. The channel thermal noise of the noise-cancelation stage (M_2 in the second stage in Fig. 2.2) was reduced by applying the noise reduction by M_{aux} of Fig. 2.3.

These techniques are now combined such that the channel thermal noise of the noise-cancelation stage, which operates now on the input-matching CG stage, is

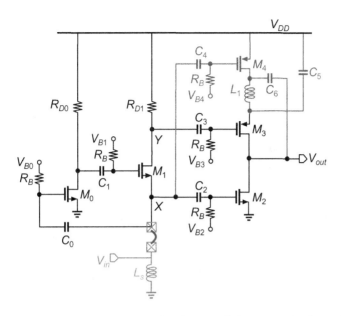

Figure 2.6 Wideband LNTA with noise-cancelation and -reduction techniques.

reduced by applying the same noise-reduction technique. Figure 2.6 shows the designed wideband LNTA. We take advantage of the CG input stage, M_1, to provide the wideband 50 Ω input matching. M_2 and M_3 of the CS stages are configured to cancel the channel thermal noise of M_1. To reuse the M_2 current and to improve the IIP3 linearity, M_3 is chosen now as a pMOS transistor. The external antenna-port inductor L_s is employed to provide a dc current path to ground and to damp the total parasitic capacitance at the input node. In this noise-reduction technique, by exploiting the current reuse, transistor M_4 is paralleled AC-wise with M_2, thus boosting its transconductance and hence decreasing its thermal noise effects. The pMOS–nMOS structure and "sweet spot" biasing are applied to improve the linearity. Moreover, the off-chip inductor L_s is on a PCB, hence its value can be fairly large, in the order of a few hundred of nH, which can resonate out all parasitics at the input node at 1.2–1.5 GHz [63].

2.3.1 Input Matching

To consider the body effect of the wideband input-matching common-gate M_1 transistor and also to simplify the relations, G_{m1} stands for $(1 + R_{D0}g_{m0})(g_{m1} + g_{mb1}) \approx (1 + R_{D0}g_{m0})g_{m1}$. Hence, the input impedance is given by:

$$Z_{in} = (R_{Ls} + sL_s) \left\| \frac{1}{sC_X} \right\| \frac{1}{G_{m1}}$$

$$= \frac{R_{Ls} + sL_s}{C_X L_s s^2 + (R_{Ls}C_X + G_{m1}L_s)s + (G_{m1}R_{Ls} + 1)},$$

(2.11)

where C_X lumps the total parasitic capacitance at node X, which is damped by L_s. Since L_s is external and connected to the antenna pin, thus not consuming any extra pads on the chip, it can be fairly large (150 nH); therefore (2.11) can be simplified to $Z_{in} = 1/(sC_X + G_{m1})$. This shows that the input matching is mainly defined by M_0 and M_1. In this case, if the size of L_s changes, for instance, from 150 to 200 nH, there will be just a barely noticeable effect on S_{11}. However, the lower limit of bandwidth (f_L) will be improved. On the other hand, if the size of L_s is decreased, its series resistance, R_{Ls}, will go down (to as low as 5 Ω) due to the limited Q-factor of L_s. This resistance is paralleled with $1/G_{m1}$ and so it lowers the equivalent input impedance. Although a new technique was described in [66] to extend the bandwidth at lower frequencies without increasing the size of L_s, here an *off-chip* inductor in parallel with the IC antenna input pin is used to realize L_s in order to save the silicon die area. Although the g_m-boost transistor, M_0, adds a bit more parasitics to the input node, it is of small size, so it does not affect the bandwidth substantially. By increasing its size from $W = 10\ \mu m$ to $20\ \mu m$, the simulated upper cutoff frequency lowers by 450 MHz, from 7.78 to 7.33 GHz.

2.3.2   Gain Analysis

The equivalent impedance seen from the drain of M_1 toward the ground is termed Z_Y and is equal to $R_{D1}||[r_{ds1} + (1/sC_X||sL_s)(1 + G_{m1}r_{ds1})]||1/sC_Y$, where R_{D1} is the load resistance of M_1, and C_Y is the total parasitic capacitance at node Y. Z_{out} determines the output impedance, which is calculated as $r_{ds2}||r_{ds3}||r_{ds4}||1/sC_{out}$, where C_{out} is the total output parasitic capacitance seen by V_{out}. Therefore, the voltage gain of this LNTA is given by:

$$A_v = -\frac{1/G_{m1}}{1/G_{m1} + R_s}(G_{m1}g_{m3}|Z_Y| + g_{m2} + g_{m4})|Z_{out}|. \qquad (2.12)$$

The design can be used either as an LNTA in an integrated current-mode RX or as a stand-alone LNA if it is externally loaded by a 50 Ω termination. In the latter, the amplifier must properly handle the intermediate network of wire-bonding inductance, pad capacitance, package parasitics, and PCB transmission lines (TLs) and components. Figure 2.7 shows the simulated output impedance of the designed LNTA, which confirms that it is suitable for the current-mode application where its output impedance is at least eight times larger than the 50 Ω load impedance [67]. In this matching network, the pad capacitance is in parallel with Z_{out}, where the equivalent impedance is in series with the wire-bond inductance. The rest of the matching network is provided on the PCB by using SMD capacitors and TLs that makes the equivalent output impedance compatible with 50 Ω.

To examine the effect of the 50 Ω load impedance of the external test equipment on the gain of this structure, (2.12) for A_v is plotted in Fig. 2.8. As expected, when

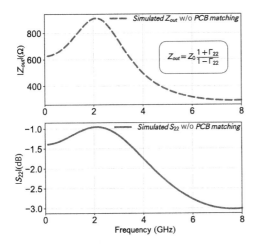

Figure 2.7 Simulated output impedance without matching network on PCB .

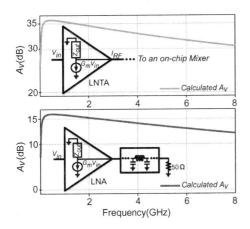

Figure 2.8 Calculated A_V from (2.12) (top) when the amplifier is self-loaded in the LNTA mode, (bottom) when the amplifier is loaded by 50 Ω through a matching network in the LNA mode.

unloaded, the voltage gain is high since Z_{out} is high.[3] When the amplifier is loaded with 50 Ω, the provided gain drops by ~20 dB.

Unfortunately, the technology scaling causes r_{ds} to be reduced. Moreover, by employing the pMOS transistors at the output node, the parasitic capacitances go up, resulting in more variation in Z_{out} at high frequencies. These are the main reasons that limit the LNTA bandwidth at high frequencies. To solve this problem, the inductive shunt-peaking and series-peaking techniques can be used. The shunt inductive peaking

---

[3] When the LNTA transconductance drives an internal on-chip mixer, its *voltage* gain will actually be very low, but it can be recovered in subsequent stages [68].

causes a resonance at the output of each stage when the gain starts to roll off at higher frequencies [66]. It is worth mentioning that L_1 also helps to dampen the parasitic capacitance at the output node. By increasing L_1 from 240 pH to 1.2 nH, the 3 dB bandwidth can be extended from 7.5 to 9 GHz. The quality factor of L_1 improves the gain only marginally. When it is increased from 5.5 to 10 ($L_1 = 440$ pH), the gain improves only by 0.1 dB.

2.3.3   Noise Analysis

As mentioned in Section 2.1, the purpose of noise-cancelation is to disassociate the input matching from the noise considerations by virtue of canceling the noise from the matching stage at the output node [49]. In this LNTA, the current noise of the input transistor flows *into* node X but *out of* node Y, causing two voltages with opposite phases. These two voltages are converted into currents by M_2 and M_3 [69]. However, the input signal appears at these two nodes at the same phase. Thus, the input signal is constructively combined at the output. The two noise voltages are calculated as $\overline{V_{nX}}^2 = Z_{in}^2 \overline{I_{n,M1}^2}$ and $\overline{V_{nY}}^2 = Z_Y^2 \overline{I_{n,M1}^2}$. Therefore, the output current noise due to the thermal noise of M_1 is as follows:

$$\overline{I_{n,out}}^2 = \overline{V_{nX}}^2 (g_{m2} + g_{m4})^2 - \overline{V_{nY}}^2 g_{m3} = 0$$

$$\Rightarrow \frac{g_{m2} + g_{m4}}{g_{m3}} = \frac{Z_Y}{Z_X}. \tag{2.13}$$

To reuse the current of M_2, M_3 is chosen as a pMOS transistor. In addition, the noise-reduction technique is applied to improve the NF without any additional power cost. In this technique, M_4 is in parallel with M_2, and hence, the transconductance of M_4 is added to that of M_2. Moreover, M_4 is selected as a pMOS transistor in order to be able to reuse the current of M_2. The consequential increase of M_2's transconductance reduces the channel thermal noise of the cancelation stage, thus avoiding any need for extra branches. Consequently, the improvement in noise figure is achieved without burning more current, as explained in Section 2.2.

The most important noise sources in this noise-cancelation scheme are the thermal noise of R_{D1} and the channel thermal noise of transistors M_2, M_3, and M_4. The noise factor of this LNA is equal to $F = 1 + F_{R_{D1}} + F_{M2} + F_{M3} + F_{M4}$, where the $F_{R_{D1}}$ term is given by the following relation:

$$F_{R_{D1}} = \frac{4kT R_{D1}(g_{m3}|Z_{out}|)^2 (Z_{o1}/(Z_{o1} + R_{D1}))^2}{4kT R_s A_v^2}$$

$$\cong \frac{R_s}{R_{D1}}, \tag{2.14}$$

where, according to Fig. 2.6, $Z_{o1} = [r_{ds1} + (R_s||1/sC_X||sL_s)(1 + G_{m1}r_{ds1})]$ and $Z_Y = R_{D1}||Z_{o1}$ when the parasitic capacitance at node Y is not considered for simplicity. A_v is the voltage gain of the LNTA, which is simplified by considering the noise-cancelation and input-matching conditions, $(g_{m2} + g_{m4})R_s = g_{m3}R_{D1}$ and $Z_{in} = R_s = 1/G_{m1}$, respectively. The other constituting terms of the noise factor F are

$$
\begin{aligned}
F_{M2} &= \frac{4kT g_{m2}|Z_{out}|^2}{4kT R_s A_v^2}\frac{\gamma}{\alpha} = \frac{4g_{m2}}{R_s(Z_Y G_{m1}g_{m3}+g_{m2}+g_{m4})^2}\frac{\gamma}{\alpha} \\
&\cong \frac{g_{m2}}{R_s(g_{m2}+g_{m4})^2}\frac{\gamma}{\alpha},
\end{aligned}
\tag{2.15}
$$

$$
\begin{aligned}
F_{M3} &= \frac{4kT g_{m3}|Z_{out}|^2}{4kT R_s A_v^2}\frac{\gamma}{\alpha} = \frac{4g_{m3}}{R_s(Z_Y G_{m1}g_{m3}+g_{m2}+g_{m4})^2}\frac{\gamma}{\alpha} \\
&\cong \frac{R_s}{|Z_Y|^2 g_{m3}}\frac{\gamma}{\alpha},
\end{aligned}
\tag{2.16}
$$

$$
\begin{aligned}
F_{M4} &= \frac{4kT g_{m4}|Z_{out}|^2}{4kT R_s A_v^2}\frac{\gamma}{\alpha} = \frac{4g_{m4}}{R_s(Z_Y G_{m1}g_{m3}+g_{m2}+g_{m4})^2}\frac{\gamma}{\alpha} \\
&\cong \frac{g_{m4}}{R_s(g_{m2}+g_{m4})^2}\frac{\gamma}{\alpha}.
\end{aligned}
\tag{2.17}
$$

By considering the noise-cancelation condition, (2.16) can be simplified as:

$$
\begin{aligned}
F_{M3} &= \frac{\gamma R_s}{\alpha|Z_Y|^2 g_{m3}} \cong \frac{\gamma R_s}{R_{D1}(g_{m2}+g_{m4})R_s\alpha} \\
&\cong \frac{\gamma}{\alpha R_{D1}(g_{m2}+g_{m4})}.
\end{aligned}
\tag{2.18}
$$

Finally, the total noise factor of the LNTA is approximately given by:

$$
F \cong 1 + \frac{R_s}{R_{D1}} + \frac{\gamma}{\alpha R_{D1}(g_{m2}+g_{m4})} + \frac{\gamma}{\alpha R_s(g_{m2}+g_{m4})},
\tag{2.19}
$$

where the fourth component is the total noise factor due to M_2 and M_4 transistors. According to (2.19), to reduce the noise contribution of R_{D1}, its value should be increased, but this is limited by the voltage drop on R_{D1}. In addition, the channel thermal noise of M_3 can be decreased by enhancing g_{m2}. As suggested by (2.19), the noise factor of M_2 is decreased since g_{m4} is added to g_{m2} without any power penalty.

The simulated relative contributions of noise sources to the total noise factor, F, at 800 MHz are shown in Fig. 2.9. The designed LNTA is compared with two other designs: (1) the CG topology shown in Fig. 2.20(a) without any noise-cancelation and -reduction techniques, and (2) the designed structure but without M_4, that is, without the noise-reduction technique. In this comparison, the LNTA with and without M_4 consumes 4.5 mW with the same-sized transistors. The size of the transistor in the CG structure is the same as the size of the CG transistor in the designed LNTA, and also its power consumption is exactly like the power consumption of the first stage in this structure, which is about 1.5 mW.

As revealed in Fig. 2.9, the CG structure (top row bars) suffers from high noise. The channel thermal noise of the main transistor, M_1, is 41% of the total noise factor. By canceling its noise, the next highest contributor is M_2. The second row (CG and NC)

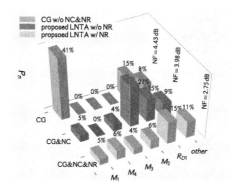

Figure 2.9 Relative contributions to the total noise factor F of various circuit components at 800 MHz for: CG structure ("CG w/o NC & NR"), designed structure without the noise-reduction technique ("LNTA w/o NR"), and the designed LNTA ("LNTA w/ NR"). *Note.* The complement to 100% is due to the 50 Ω antenna-terminal thermal source.

shows that the thermal noise contribution of the main transistor, M_1, is reduced to 5%, whereas the thermal noise of the cancelation transistor, M_2, is added with a contribution of almost 27%. By using both the noise-reduction (NR) and noise-cancelation (NC) techniques (bottom row bars in Fig. 2.9), the thermal noise contribution of M_2 is decreased to 6%, thus improving the system noise performance. The thermal noise of R_{D1} is now dominant. According to the second term of (2.19), to reduce the noise effect of R_{D1}, its value should be increased. However, the value of R_{D1} is limited by the supply voltage of the first stage, which should be at a certain level in order to provide the input matching. Therefore, to further improve the noise performance, a g_m-boosting technique by means of M_0 is introduced. This way, the amount of current of the first stage decreases as well as the voltage drop on M_1. Consequently, the value of R_{D1} can be increased, leading to the decrease of NF. The g_m-boosting stage of M_0 boosts the g_{m1} of the input stage, $G_{m1} = (1 + g_{m0}R_{D0})g_{m1}$, so the input matching can be provided with less current.

2.3.4   Linearity and Stability

Since the nonlinearity of a CS configured transistor is worse than that of the CG, the pMOS–nMOS structure placed at the output stage turns out to also improve the second- and third-order nonlinearities. By using a power series, the total output current of the pMOS and nMOS transistors in the complementary connection is equal to $i_{ds_{tot}} = i_{ds_P} + i_{ds_N} = (g_{mN} + g_{mP})(v_g - v_s) + (g'_{mN} - g'_{mP})(v_g - v_s)^2 + (g''_{mN} + g''_{mP})$ $(v_g - v_s)^3$, where g_m, g'_m, and g''_m are the first-, second-, and third-order derivatives of the transistor's composite (large-signal) drain-source current, i_{ds}, with respect to its composite gate-source voltage, v_{gs}. Since the AC input signal for the pMOS and nMOS transistors is out of phase, the total transconductance increases while the total second nonlinear term, $g'_{mN} - g'_{mP}$, decreases [61]. Figure 2.10 shows that by applying

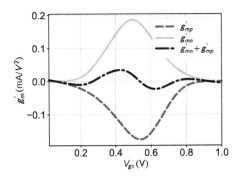

Figure 2.10 Second-order nonlinear components of g_m of pMOS and nMOS transistors.

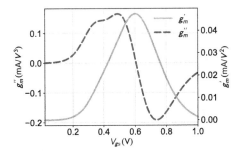

Figure 2.11 Second- and third-order derivatives of the drain-source dc current, i_{ds}, with respect to V_{gs} of M_1.

the noise-reduction technique, the pMOS and nMOS transistors, M_2 and M_4, in fact are like a complementary circuit in the output stage, which causes the second-order nonlinear components, g'_{mP} and g'_{mN}, to neutralize each other within the range of the bias voltage. As a result, the second-order nonlinear term is attenuated, and since the second-order nonlinear current can be mixed with the input by the feedback path through c_{gd} [61], both IIP2 and IIP3 are significantly improved. However, in this design, the pMOS–nMOS pair is not considered to be biased at the exact point where $g'_{mn} + g'_{mp} = 0$. The measured linearity variation due to different voltage biases of the pMOS–nMOS pair is less than 2 dB. It is worth mentioning that the linearity performance deteriorates a bit (<2 dB) by adding M_4 due to lowering of the available voltage swing in the output stage.

Consequently, to improve the linearity of the CG transistor, it is biased in a "sweet spot." According to Fig. 2.11, at the right bias voltage at which the third-order nonlinear component of the CG transistor, g''_m, is equal to zero, the IIP3 of the CG structure can be improved. It is worth mentioning that by modeling the circuit's nonlinearity via Volterra series, it can be shown that the parasitic capacitance can also affect the second- or third-order nonlinearity cancellation based on the "sweet spot." Although the sweet spot could be a bit shifted with frequency, it will be demonstrated in Section 2.3.5 that the variation of the measured IIP3 is within 1 dB across the entire bandwidth.

The most important drawback of the sweet-spot technique is its sensitivity to the process corners [70], which might require process calibration. Another option could be a constant-g_m biasing circuit. Once the sweet spot has been calibrated for the process, the LNA is quite insensitive to temperature and voltage variations. The reason is that M_1, located in the first stage, is mainly used for input matching, so its effective gain is small, and thus its linearity contribution is not dominant and the signal provided to the second stage is still small. In other words, it is biased mainly to provide the required g_m for the input matching.

To examine the stability of the LNTA with an arbitrary source and load impedances, the Stern stability factor defined in (2.20) is often utilized [71]:

$$K = \frac{1 + |\Delta|^2 - |S_{11}|^2 - |S_{22}|^2}{2|S_{21}||S_{12}|}, \tag{2.20}$$

where $\Delta = S_{11}S_{22} - S_{12}S_{21}$; S_{11}, S_{22}, S_{21}, and S_{12} are the input return loss, output return loss, forward gain, and reverse gain, respectively. If $K > 1$ and $\Delta < 1$, then the circuit is unconditionally stable [71]. According to (2.20), the stability of the circuit is improved by maximizing the reverse isolation.

2.3.5   Measurement Results

The designed wideband LNTA, whose chip micrograph is shown in Fig. 2.12, is fabricated in TSMC 28 nm bulk LP CMOS. Although this amplifier is specifically designed to drive a mainly capacitive load of integrated mixers as a high-impedance transconductor (thus, LNTA), it is also capable of driving heavy external resistive loads. Hence, it can also function as an LNA with 50 Ω input and output ports. To avoid adding an extra test buffer for driving the output port, which would need to be separately characterized, all the performance and power consumption measurements are with the external load of 50 Ω. By carefully sizing the transistors and using the noise-cancelation and -reduction techniques (with current reuse), this amplifier operates at a 1 V power supply with a power dissipation of 4.5 mW, while achieving a remarkably high and flat small-signal gain and a very low noise figure in the whole wide bandwidth.

Measured S-parameters of the LNTA are shown in Fig. 2.13, which illustrates the best input return loss of around 2.7 GHz (i.e., the input impedance is matched at

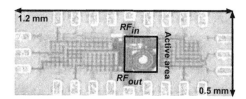

Figure 2.12 Microchip photograph.

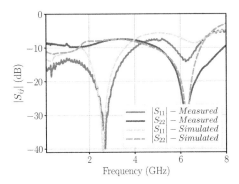

Figure 2.13 Measured and simulated input/output return loss.

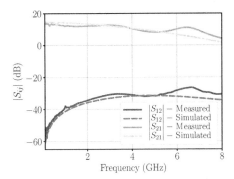

Figure 2.14 Measured and simulated gain and isolation.

this frequency). Although the input return loss gets worse away from this point, the wideband input-matching feature is well controlled as $S_{11} < -10\,dB$ in the whole bandwidth. The measurement results are well matched with the simulations. Figure 2.14 shows the power gain that varies between 12 and 15.2 dB in the range of 20 MHz to 4.5 GHz. By adding transistor M_4, the second-stage transconductance in the presented LNA increases, resulting in more power gain, which is also expected from (2.12). Since the drains of three transistors, M_2, M_3, M_4, are connected to the output node, the total parasitic capacitance at this node increases. Hence, the $-3\,dB$ bandwidth of the LNTA is partially decreased. However, L_1 helps to dampen the parasitic capacitance at the output node and compensate for the reduction in bandwidth. Unfortunately, the measurement results of the bandwidth fall short mainly because of the larger wire-bonding inductance and parasitic capacitance of the pad and PCB traces affecting the *dominant* pole at the external output port. It is worth mentioning that this issue is irrelevant in integrated receivers or if the LNA is followed by an integrated mixer on the same die.

The measured NF of the LNTA is superimposed on the simulated NF in Fig. 2.15. It varies from 2.09 dB to 3.2 dB in the 4.4 GHz bandwidth. A two-tone RF signal

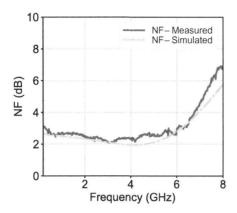

Figure 2.15 Measured and simulated NF.

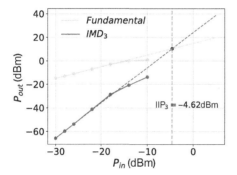

Figure 2.16 Measured IIP3 at maximum gain.

at 500 MHz, 2 GHz, and 4 GHz (i.e., at the beginning, middle, and end of the band, respectively) is used to measure the wideband linearity performance. To examine the flatness of linearity, various two-tone spacings of 2.5, 10, 50, and 100 MHz are applied but, as expected, exhibited no difference in performance. As shown in Fig. 2.16, the measured IIP3 at 500 MHz with 10 MHz spacing, where maximum gain is achieved, is −4.63 dBm, which is the minimum IIP3 in the entire bandwidth. Figure 2.17 shows the measured IIP2 and IIP3 versus frequency. Note that in integrated designs there is always a dc-blocking capacitor between the LNA and a passive mixer, so the dc will be blocked and low-frequency IM2 products will be heavily attenuated. Without the 50 Ω load, the simulations show the linearity of −8.7 dBm at the gain of 35.2 dB.

Finally, to verify the stability, the Stern stability factor (2.20), K, with Δ are plotted in Fig. 2.18 based on the measured data. As evident, the LNA is stable over the whole bandwidth, as $K > 1$ and $\Delta < 1$.

To compare this LNTA with prior-art architectures and to emphasize the capabilities of reaching lower frequencies in this wideband design, the following

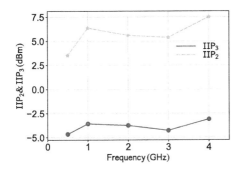

Figure 2.17 Measured IIP3 and IIP2 versus frequency.

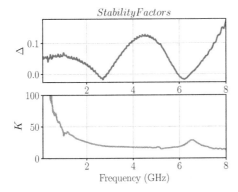

Figure 2.18 Measured stability factors K and Δ.

figures-of-merit (FoM$_2$ and FoM$_3$) are defined based on the original FoM (termed here FoM$_1$) introduced in [54], and the results are summarized in Table 2.1.

$$\text{FoM}_1 = \frac{Gain_{av}[\text{abs}] \times (f_H - f_L)[\text{GHz}]}{(F_{av} - 1) \times P_{dc}[\text{mW}]}, \tag{2.21}$$

$$\text{FoM}_2 = \frac{Gain_{av}[\text{abs}] \times (f_H - f_L)[\text{GHz}]}{(F_{av} - 1) \times f_L[\text{GHz}] \times P_{dc}[\text{mW}]}, \tag{2.22}$$

$$\text{FoM}_3 = \frac{Gain_{av}[\text{abs}] \times (f_H - f_L)[\text{GHz}] \times \text{IIP3}[\text{mW}]}{(F_{av} - 1) \times f_L[\text{GHz}] \times P_{dc}[\text{mW}]}, \tag{2.23}$$

where F_{av} is the average noise factor, $Gain_{av}$ is the average power gain over the 3 dB frequency range f_L to f_H, and P_{dc} is the power consumption. Even without any extra output buffer to mitigate the loading effects of the external 50 Ω termination, the introduced LNTA provides a very low noise figure and has competitive power consumption for the ultrawide bandwidth (4.48 GHz), which is achieved by virtue of using both noise-reduction and -cancellation techniques. Moreover, the circuit has a

Table 2.1. Summary and comparison with state-of-the-art wideband LNAs

	CMOS tech. [nm]	BW [GHz]	S_{11} [dB]	S_{21} [dB]	IIP3 [dBm]	NF [dB]	V_{DD} [V]	Power [mW]	Active area [mm²]	Can Noise cancel. used?	Can drive extern. 50Ω?	FoM$_1$	FoM$_2$	FoM$_3$
This work	**28**	**0.02~4.5**	**≤ −10**	**15.2**	**−4.62 ~ −3.53**	**2.09~3.2^d**	**1**	**4.5**	**0.03**	**Yes**	**Yes**	**7.76**	**327**	**172.6**
[72] TCASI'20	65	0.05~1.3	≤ −10	27.5	−4 ~ −1	2.3~3	1	5.7	0.046	Yes	No^b	6.18	123.7	6.95
[73] TMTT'20	65	1~20	≤ −10	12.8	1~5.8	3.3~5.3	1.6	20.3	0.096	Yes	No^b	2.4	2.4	5.28
[74] TCASI'19	65	0.05~1	≤ −10	30	−10 ~ −2.5	2.3~3.3	2.2	19.8	0.0448	Yes	No^b	1.6	33.51	26.6
[75] TCASII'19	65	0.4~2.2	–	16.4	−5	2~2.5	1.2	29	0.16	Yes	Yes	0.6	1.5	0.47
[76] TMTT'19	65	0.3~4.4	–	26.7	−14.2	3~4.4	1	13.7	0.009	No	No^b	4.8	16	0.6
[77] TCASII'19	65	0.5~7	–	16.8	−4.5	2.87~3.77	1.2	11.3	0.044	Yes	No^b	3.5	7	2.48
[78] TCASI'18	65	0.2~2.7	< −5	21.2	−2	3~3.5	1.2	0.96	0.05	Yes	No^b	26.85	134.2	84.7
[79] TCASII'18	180	2~5	–	13	−9.5	6~8	1.8	1.8	0.72	Yes	No^b	1.85	0.93	0.1
[80] JSSC'17	180	0.1~2	–	17.5	10.6	2.9~3.5	2.2	21.3	0.63	Yes	Yes^a	0.6	6.14	70.5
[81] JSSC'2016	130	0.6~4.2	< −10	14	−10	4~9	0.5	0.25	0.39	No	No^b	20.8	34.7	3.46
[82] TMTT'16	130	0.1~2.2	–	12.3	−11.5 ~ −9.5	4.9~6	1	0.4	0.0052	Yes	No^b	8.6	86.2	9.68
[83] TCASI'14	90	3.5~9.25	≤ −8	15	−16.3 ~ −12	2.4	0.8	9.6	0.56	No	–	4.5	1.3	0.0823
[84] MWL'14	180	0.02~1.4	< −10	16.4	−13.3 ~ −9	3~4.7	1.8	12.8	0.04	No	No^c	0.5	24.96	3.1
[85] JSSC'13	65	0.1~10	≤ −11	24	−15 ~ −12	2.59~4.92	1.2	8.64	0.012	No	–	15.29	152.9	9.65
[86] TCASI'12	65	0.1~5.1	–	10.7	~6	2.9~5.4	1	6	0.03	Yes	–	1.75	17.52	69.7
[87] JSSC'12	130	0.1~2.7	–	20	−12	4	1.2	1.32	0.007	No	No^c	13	130.3	8.2

^{a} Single-ened load is 100 Ω. ^{b} Uses addt'l on-chip measurement buffer.
^{c} Needs addt'l external measurement buffer. ^{d} Measured over 100 MHz–6.5 GHz

competitive linearity and quite high power gain versus the other leading designs. As shown in the comparative landscape in Fig. 2.19, this design achieves the best FoM among the recent state-of-the-art LNAs. Moreover, one of the main advantages of this architecture compared to prior reports is that it provides a high impedance at its output, which makes it suitable to drive an integrated passive mixer in a modern receiver. Despite the use of the additional on-chip (0.3 nH) inductor, the area still remains very competitive.

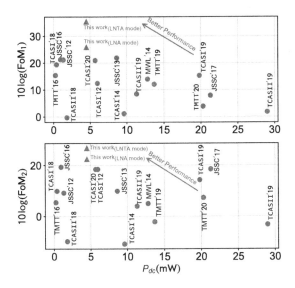

Figure 2.19 FoM landscape of high-performance LNAs.
Note. FoM$_{2,3}$ of our work in the LNTA mode (assuming a hypothetical test buffer) is based on a combination of measurements and simulations.

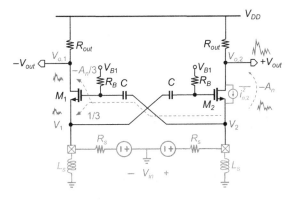

Figure 2.20 Cross coupled common-gate LNA [88] with reduced required input-matching transconductance and partial noise input noise cancelation at differential output.

2.4 TwoFold Cross-Coupled Wideband LNTA architecture

In this section, another technique is introduced. It improves the noise performance of the CG structure while providing high gain. In Section 2.4.1, it will be shown in more detail how a twofold cross-coupled LNTA can be designed.

2.4.1 Cross-Coupled Common-Gate LNA

As the first basic LNA structure, cross-coupled common-gate LNA [88] is shown in Fig. 2.20. To have input source impedance (R_s) matching in this structure, input transistors ($M_{a,n}$ and $M_{a,p}$, here after M_a) should have a transconductance of:

$$g_m = \frac{1}{2R_s}.$$ (2.24)

This is half the required g_m for a common-gate LNA without the cross-coupling, which saves power consumption. To analyze the noise of this structure, initially the noise of $M_{a,p}$ (indicated in Fig. 2.20) is only considered in the right branch. First, by writing KCL at $V_{a,n}$, we have:

$$\frac{V_1}{R_s} = (V_2 - V_1)g_{m1}.$$ (2.25)

Substituting (2.24) in (2.25) gives:

$$V_1 = \frac{1}{3}V_2.$$ (2.26)

Then, by writing KCL at V_2, we have:

$$\frac{V_1}{R_s} = (V_2 - V_1)g_1 + i_{n2},$$ (2.27)

where $\overline{i_2^2}$ is a spectral density of M_2 noise current. By replacing (2.26) in (2.27) and considering the input-matching condition, it is simplified to:

$$V_2 = \frac{3}{4}i_{n2}R_s$$ (2.28)

It could be easily shown that the M_2 noise appears the V_{o2} with $-A_n$ times V_1, where A_n is given by:

$$A_n = \frac{R_{out}}{R_s}$$ (2.29)

Therefore, using (2.26), (2.28), and (2.29), the effect of $\overline{i_{n2}^2}$ on V_2 and V_1 is found:

$$A_n$$ (2.30)

In addition to M_2 noise, R_{out} noise $(\overline{v_{n,R_{out}}^2})$ also goes to the output. Note that the noise of the left branch is also the same amount. Moreover, all noise sources are uncorrelated. While the input is matched, the signal gain from input to output is calculated:

$$A_v = \frac{V_{out,d}}{V_{in,d}} = \frac{V_{o,d}}{V_{a,d}} = g_{m1} \times 2R_s = \frac{R_{out}}{R_s}.$$ (2.31)

Now, noise factor (F) (i.e. noise figure (NF) is equal to $10\log F$) of the LNA can be found by referring the effect of different noise sources from the output $V_{o,diff}$ to the input $V_{a,diff}$, and normalizing to the noise of input source (R_s):

$$F = 1 + \frac{\frac{2\times((1/2)i_{n2}R_{out})^2}{A_v^2} + \frac{2\times v_{n,R_{out}}^2}{A_v^2}}{2 \times (1/2)^2 \times v_{n,R_s}^2} = 1 + \frac{\frac{2\times(1/2)^2 4kT\gamma_a g_{m,a}\times(R_{out})^2}{(R_{out}/R_s)^2} + \frac{2\times 4kT R_{out}}{(R_{out}/R_s)^2}}{2 \times (1/2)^2 \times 4kT R_s}.$$

(2.32)

The $2\times$ multiply in the numerators and the denominator is to account for the noise of both branches deferentially. Furthermore, the $1/2$ coefficient in the denominator is because of the gain of $1/2$ from the input source to LNA input made by input matching. M_2 noise, $\overline{i_{n2}^2}$, is considered $4kT\gamma g_m$, where k, T, and γ are Boltzmann constant, absolute temperature, and MOS noise excess factor, respectively. After simplification and considering the input-matching condition, (2.32) can be written as:

$$F = 1 + \frac{\gamma_a}{2} + \frac{4}{R_{out}/R_s}. \tag{2.33}$$

The second term is the noise contribution of M_a transistors that is reduced to half due to the cross-coupling and less required $g_{m,a}$ for input matching. Supposing $\gamma_a \sim 1$ in nano-scale CMOS and 20 dB voltage gain, NF of this LNA is limited to ≥ 2.8 dB.

2.5 Basic Common-Gate Noise-Canceling Structure

The common-gate-source-follower noise-canceling LNA structure [90] shown in Fig. 2.21 is able to cancel noise of the input-matching transistors. In this LNA, R_{out} is replaced with two transistors, M_3, that transfer signal and noise of the input nodes (V_1) to the outputs (V_3). Input signal goes to the output from two paths, through M_1 and M_3 that are added in phase on V_3 node. However, M_1 noise reaches to the output from the two paths antiphase, and therefore can be reduced or canceled out.

Input matching in this LNA is achieved by:

$$g_{m,a} = \frac{1}{R_s}. \tag{2.34}$$

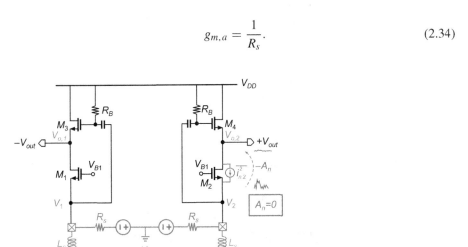

Figure 2.21 Common-gate-source-follower noise-canceling LNA [89] with noise cancelation of the input-matching transistors.

Substituting (2.33) in (2.34) gives: First, the noise of M_2 is only considered in the right branch. By writing KCL at V_2, we have:

$$\frac{V_2}{R_s} = -V_2 g_{m2} + i_{n2}.$$ (2.35)

Substituting (2.34) in (2.35) gives:

$$V_2 = -\frac{1}{2} R_s i_{n2}.$$ (2.36)

Then, KCL at V_4 node gives:

$$i_{n2} - V_2 g_{m2} = (V_2 - V_4) g_{m4}.$$ (2.37)

By replacing (2.34) in (2.36) gives:

$$V_2 = \frac{1}{2} i_{n2} \left(R_s - \frac{1}{g_{m4}} \right).$$ (2.38)

Here M_2 noise transfer ration (A_n) from V_2 to V_4 is found from (2.36) and (2.38):

$$A_n = -\frac{V_4}{V_2} = \frac{1}{g_{m4} R_s} - 1.$$ (2.39)

To have a perfect noise cancelation of M_2, A_n should become zero. Hence we have:

$$A_n = 0 \Rightarrow g_{m4} = \frac{1}{R_s}.$$ (2.40)

With this condition, the amount of M_2 noise current that directly goes to the output ($V_{o,2}$) is the same amount transferred from M_4, but with the opposite direction. In this way, only M_4 noise is left in the right branch so that its output noise voltage is trivially:

$$\overline{v_{n4}^2} = \frac{\overline{i_{n4}^2}}{g_{m4}^2}.$$ (2.41)

Considering the input-matching (2.34) condition, the signal voltage gain from the two paths is derived:

$$A_v = \frac{V_{out}}{V_{in}} = \frac{g_{m,z}}{g_{m4} + 1} = \frac{a}{g_{m4} R_s} + 1,$$ (2.42)

where the first and second terms are paths through M_2 and M_4, respectively. Note that the signal gain (A_v) in (2.42) is different than the M_2 noise ratio (A_n) in (2.39). Moreover, accounting for M_2 noise-cancelation (2.40) condition, (2.32) is reduced to:

$$A_n = 0 \Rightarrow A_v + 2.$$ (2.43)

Therefore, the noise-cancelation condition forces a fixed voltage gain in this LNA. Noise factor of this LNA is found by considering the only remaining noise source in (2.41):

$$F = 1 + \frac{\frac{4\overline{i_{n4}^2}/g_{m4}^2}{A_v^2}}{v_{n,R_s}^2} = 1 + \frac{4kT\gamma_4}{4kT R_s g_{m4}} = 1 + \gamma_4. \tag{2.44}$$

Although the noise of the input matching is eliminated compared to (2.33), M_4 noise still imposes a high NF. Supposing $\gamma_4 \approx 1$, NF of this LNA is limited to >3 dB while providing only 6 dB voltage gain.

2.5.1 Noise-Canceling Structure with Input Cross-Coupling

The first step to realize this LNTA structure is combining the two cross-coupled common-gate LNA and the basic common-gate noise-canceling structures (Fig. 2.22). In this structure, the input, matching condition is the same as (2.24), $g_{m1} = 1/(2R_s)$. Moreover, similar to the cross-coupled common-gate LNA in Fig. 2.20, M_1 noise appears on the input nodes:

$$\begin{cases} V_2 = \frac{3}{4}i_{n2}R_s, \\ V_1 = \frac{1}{3}V_2, \end{cases}$$

KCL at $V_{b,p}$ gives us:

$$i_{n2} + g_{m2}(V_1 - V_2) = (V_2 - V_{o,2})g_{m4}. \tag{2.45}$$

By substituting (2.45), M_2 noise on the output node is found:

$$V_{o,2} = \frac{3}{4}i_{n2}\left(R_s - \frac{1}{g_{m4}}\right). \tag{2.46}$$

Therefore, this structure also has the same noise ratio as in (2.39):

$$A_v = -\frac{V_{o,2}}{V_2} = \frac{1}{g_{m4}R_s} - 1. \tag{2.47}$$

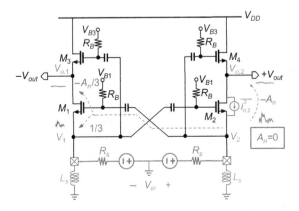

Figure 2.22 Common-gate-source-follower noise-canceling LNA with input cross-coupling.

Again here, $g_{m4} = 1/R_s$ leads to complete M_2 noise cancelation. Signal voltage gain is similarly derived the same as (2.42) that can be alternately written as:

$$A_v = -A_n + 2. \qquad (2.48)$$

This LNA structure still has the noise figure of (2.44). However, it requires less power consumption to provide the input matching. In addition to the problem of a high NF limitation, this structure still does not have the freedom to set the voltage gain (A_v).

2.5.2   Adding g_m-Cell

As the final design goal is to have an LNTA, an inverter-based g_m-stage is added at the output of the LNA core of Fig. 2.23. However, instead of connecting both NMOS and PMOS to $V_{o,1}$ node, one is connected to $V_{o,1}$ and the other one to V_1 (see Fig. 2.23, LNTA #1). In this way, M_1 noise cancellation can be done for an arbitrary voltage gain (A_v) by exploiting an available degree-of-freedom in the g_m-stage, where signals from V_2 and $V_{o,2}$ to out has a voltage-to-current gain of $g_{m,\alpha}$ and $g_{m,\beta}$, respectively. Then, instead of setting $g_{m8} = 1/R_s$ to cancel M_1 noise in the LNA core itself, a lower g_{m8} can be used. Consequently, A_v is increased that lowers input-referred noise of M_8 and the g_m-stage. To cancel M_1 noise at the output of g_m-stage, we have:

$$g_{m4} V_2 + g_{m6} V_{o,2} = 0. \qquad (2.49)$$

Substituting (2.47) into (2.49) gives:

$$g_{m4} V_2 + g_{m6}(-A_v V_2) = 0 \Rightarrow \frac{g_{m4}}{g_{m6}} = A_v. \qquad (2.50)$$

With this condition satisfied, the noise of M_2 does not appear at the output node. As calculated in (2.45), this noise appears on each corresponding node at the left side of LNA core with a gain of 1/3 Fig. 2.23. Consequently, in the same way, it is also canceled at output.

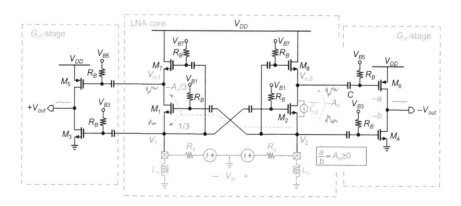

Figure 2.23 Noise-canceling LNTA (structure 1). Noise of the input-matching transistors are cancelled out with an arbitrary gain of LNA core.

Total gain of the LNTA from input to output is provided by two paths: through V_1 and V_2 nodes. Using (2.48) and (2.50), the total single-ended transconductance is derived:

$$g_{m,total} = -(g_{m4} + A_v g_{m6}) = -2g_{m6}(1 + A_n). \quad (2.51)$$

Total noise factor of the LNTA is calculated by input-referring noise of M_4, M_6, and M_8 from the output to the input:

$$F = 1 + \frac{\frac{(\overline{i_{n8}^2}/g_{m8}^2) \times g_{m6}^2}{g_{m,total}^2} + \frac{\overline{i_{n4}^2} + \overline{i_{n6}^2}}{g_{m,total}^2}}{(1/2)^2 4kT R_s} = 1 + \frac{\frac{(4kT\gamma_b/g_{m8}^2) \times g_{m6}^2}{g_{m,total}^2} + \frac{4kT(\gamma_a g_{m4} + \gamma_\beta g_{m6})}{g_{m,total}^2}}{kT R_s} \quad (2.52)$$

After assuming $\gamma_a = \gamma_b$ and replacing (2.51), the noise factor is simplified to:

$$F = 1 + \frac{\gamma_b}{(1 + A_n)} + \frac{2\gamma_a}{g_{m,total} R_s}. \quad (2.53)$$

The second term is due to the noise of M_b that is reduced $1 + A_n$ times by the signal gain from other paths. The third term is the total noise contribution of the g_m-stage that is reduced two times by the total multipath gain provided in the LNA core (compared with NF of a stand-alone g_m-stage).

2.5.3 Final Structure with a Second Noise Cancelation

The fully differential LNA, shown in Fig. 2.24, realizes the twofold noise cancelation to further improve the circuit's noise performance. The core of this structure consists of the cross-coupled common-gate (CG) topology, $M_1(M_2)$, which is used for implementing the broadband input matching. The cross-coupled structure helps to provide input matching with half of the required g_m, thus leading to a reduced power

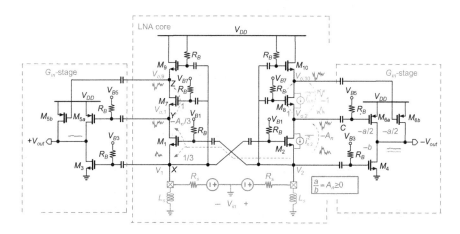

Figure 2.24 Wideband noise-canceling LNTA (structure 2). Noise-cancelation mechanisms of M_2 and M_8 is shown that is cancelled at the output. Noise of M_8 goes to the output with half gain, $\beta/2$.

consumption. The source-follower transistors, M_7 and M_9 (M_8, M_{10}), complete the core of the twofolded noise cancelations. The common-source (CS) transistors, M_{5a} and M_3 (M_{6a}, M_4), cancel the channel thermal noise of M_1 (M_2). The second noise-cancelation stage helps to cancel the channel thermal noise of M_7 (M_8) by using M_{5a} and M_{5b} (M_{6a}, M_{6b}).

As was derived in (2.53), the noise of g_m-stage can be reduced by increasing the total g_m. Moreover, M_2 noise can be lowered by increasing the voltage gain of the LNA core. However, to reach a very low NF in the range of $1 - 2$ dB, excess increase of voltage gain degrades linearity of the g_m-stage.

To achieve this with a reasonable LNA core gain, M_4 noise in Fig. 2.23 is also reduced substantially, in addition to complete cancelation of M_2 noise. Instead of using only one M_4 on each side of the LNA core, two identical transistors M_8 and M_{10} are stacked (see Fig. 2.24, LNTA [1]). Although the amplified input signal appears in phase with the same gain, noise of M_8, (shown in Fig. 2.24), appears antiphase on $V_{o,2}$ and $V_{o,10}$. Then, splitting $M_{6,n}$ into two half transistors ($M_{6,a}$ and $M_{6,b}$) cancels noise of $M_{b1,p}$ at the output via the introduced noise splitting technique. This way, only noise of M_{10} contributes to the output but with a reduced gain of $g_{m6}/2$ instead of previously g_{m6} in (2.52). The noise splitting technique can be further utilized to cancel noise of M_{10} at the cost of a higher LNA core supply voltage.

2.5.4   Input Matching

In the twofold noise-canceling architecture, the common-gate (CG) stage is used to provide wideband input matching, which is roughly equal to $Z_{in} = 1/g_{m1}$ at low frequencies. Since this architecture is fully differential, the gates of the input transistors, M_1 and M_2, are cross-coupled to save power consumption. In doing so, the total g_m needed to provide the 50 Ω input matching is cut in half, that is, to 10 mS. The equivalent input impedance seen at the input node, X, is calculated as:

$$Z_{in} = (R_{Ls} + sL_s) \left\| \left\| \frac{1}{sC_X} \right\| \right\| \frac{1}{2(g_{m1} + g_{mb1})}$$

$$= \left(C_X L_s s^2 + (R_{Ls} C_X + 2(g_{m1} + g_{mb1})L_s)s \right.$$

$$\left. + (2(g_{m1} + g_{mb1})R_{Ls} + 1/2) \right)^{-1} \times \left(R_{Ls} + sL_s \right) \qquad (2.54)$$

where g_{mb1} represents the body effect of transistor M_1, R_{Ls} is a series resistance of the off-chip inductor L_s due to its finite quality factor, and C_X is the lumped parasitic capacitances at the input node X.

2.5.5   Gain Analysis

As indicated in Fig. 2.24, the input signal reaches the output node via four parallel paths. In the main path, path 1, the input signal is amplified by M_1 and M_{5a} to get to the output node. Paths 3 and 4 use M_7, M_{5a} and M_9, M_{5b}, respectively, to provide the boosted input signal at the output. Finally, the output voltage provided by path 2 is amplified by M_3.

As shown in Section 2.5.4, Z_{in} is expressed by (2.54). Z_Y and Z_Z are total impedances to ground seen at nodes Y and Z, and defined as $r_{ds1} \| (sC_Y)^{-1} \| (g_{m7})^{-1}$ and $r_{ds7} \| (sC_Z)^{-1} \| (g_{m9})^{-1}$, respectively, where C_Y and C_Z are total parasitic capacitances to ground at nodes Y and Z. Finally, Z_{out} is a total impedance seen at the output node and is equal to $r_{ds3} \| (sC_{out})^{-1} \| r_{ds5}$ where C_{out} is the total parasitic capacitance at the output. Then, the LNA gain, by considering the effects of all parasitic capacitances, is equal to:

$$A_v = \frac{Z_{in}}{Z_{in} + R_s}(2g_{m1}g_{m5}Z_Y + g_{m5a} + g_{m5b}$$
$$+ g_{m3} + g_{m1}g_{m5b}g_{m7}Z_Z Z_Y)Z_{out}. \tag{2.55}$$

According to (2.55), by increasing the transconductances g_{m1}, g_{m3}, and g_{m5}, higher gain can be achieved, but the total power consumption will be increased. Moreover, by increasing the size of M_3 and M_5 for the sake of increasing the gain, the total parasitic capacitance at the output node increases, thus leading to reduction in the $-3\,$dB bandwidth of the designed architecture.

To consider the body effect of M_1 and also to simplify the formulation, in all following equations g_{m1} stands for $g_{m1} + g_{mb1}$. To simplify (2.55), it is considered $Z_Y = 1/g_{m7}$, $Z_Z = 1/g_{m9}$, and $g_{m5a} = g_{m5b}$. By applying constraints from the calculated noise-cancelation equations in Section 2.5.6, that is, (2.3) and (2.10), the gain of the circuit can be simplified to:

$$A_v = \frac{3}{2}(g_{m5} + g_{m3})Z_{out}. \tag{2.56}$$

As explained previously, the second-stage transistors, M_3 and M_5, are the key contributions to the total gain of the circuit. Now, the total gain of LNTA from input to output is provided by three paths: through V_a, V_{b1}, and V_{b2} nodes. However, the same equation in (2.51) is still valid here (the signal voltage gain from V_{b1} to V_{b2} is 1).

2.5.6 Noise Analysis

The half-circuit of the LNA, shown in Fig. 2.25(a), is examined for its noise performance. All parasitic capacitances, including C_X, C_Y, and C_Z associated with nodes X, Y, and Z, are considered for bandwidth limitations. The channel thermal noise of M_1 and M_7 generates two antiphase noise voltages at their respective drain and source ports. By utilizing the double noise-cancelation technique, their effects are canceled at the output node, ultimately leading to the enhancement of the LNA noise performance. Moreover, the channel thermal noise of M_9 is also partially canceled through the noise-cancelation mechanism. This further reduces its contribution in the total output noise. As shown in Fig. 2.25(a), by canceling the noise of $M_{1,7}$ and reducing the noise

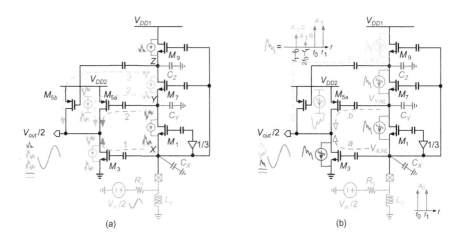

Figure 2.25 Half-circuit of twofold noise-cancelation LNA: (a) noise contributions of each stage, (b) nonlinearity contributions of each stage.

of M_9, the channel thermal noise of the remaining transistors, M_3 and M_5, is now dominant in the total noise factor at the output node.

It is worth mentioning that M_7 can be used to intrinsically cancel the noise of M_1. However, it limits the available gain of the first stage. According to Fig. 2.25(a), the two noise voltages generated by the thermal noise of M_1 at nodes X and Y are amplified through path 1 and path 2 and added together at the output node. Since the amplitude voltage noise sources at nodes X and Y are not equal, paths 1 and 2 should provide different amplification factors such that their summation becomes zero at the output node, thus leading to the first noise-cancelation condition expressed in the following equation:

$$\frac{g_{m3}}{g_{m5}} = \frac{g_{m1,T}}{g_{m7}} - 1. \tag{2.57}$$

The channel thermal noise of M_7 also generates two voltage sources at nodes Y and Z. By passing these voltages through paths 2 and 3 toward the output, they are amplified and added up at the output nodes. Since the equivalent impedances to the ground seen at nodes Y and Z are equal, the two generated voltage noises have the same amplitude at these nodes. By providing the same amplification factors through path 3 and 4, the effective contribution of the noise of M_7 becomes zero at the output node, which defines the second noise criterion brought by (2.58).

$$\frac{g_{m5b}}{g_{m9}} = \frac{g_{m5a}}{g_{m7}}. \tag{2.58}$$

By canceling the channel thermal noise of M_1 and M_7, the total noise factor of the half circuit of Fig. 2.25(a) is equal to the summation of the noise factors of M_3, M_5, and M_9, which can be derived as in (2.59):

$$\frac{F_T}{2} = 1 + \frac{1}{9R_s g_{m9}(1 + g_{m3}/g_{m5})^2} \frac{\gamma}{\alpha}$$

$$+ \frac{4}{9R_s g_{m3}(1 + g_{m5}/g_{m3})^2} \frac{\gamma}{\alpha}$$

$$+ \frac{4}{9R_s g_{m5}(1 + g_{m3}/g_{m5})^2} \frac{\gamma}{\alpha} \qquad (2.59)$$

Total new noise figure of the LNTA is derived by referring the noise contribution of M_{b2}, M_α, and M_β from the output to the input as it is derived in (2.52). According to (2.52), the second term is due to the noise of M_{b2} that is substantially reduced four times by the noise-splitting technique, and $1 + A_n$ times by signal gain from other paths.

Figure 2.26(a) plots the noise figure of the LNTA with and without the noise-splitting technique. A noise figure of even sub-1 dB could be achieved by increasing $g_{m,total}$ and A_n values. Simulated noise figure and g_m of our implementation with a

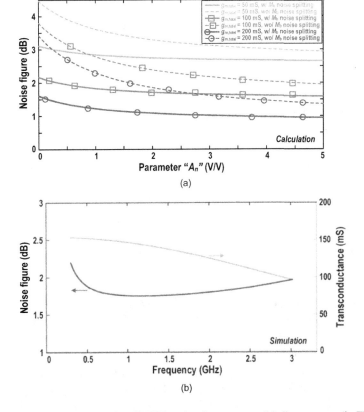

Figure 2.26 (a) Calculated LNTA noise figure versus "A_n" parameter (in Fig. 2.24), and (b) simulated noise figure and total g_m versus frequency, with $S_{11} < -10$ across the range. Note that LNA core gain is $A_v = A_n + 2$.

target of 120 mS are shown in Fig. 2.26(b). The parameter A_n is chosen to be about 1 (LNA core gain about 10 dB) in this design to have a balance between NF and IIP3. The covered LNTA frequency range is wideband: measured from 300 MHz up to 3 GHz. The cores of LNA and G_m-stages drain totally 3 mA and 10 mA from 2 V and 1.2 V power supplies, respectively.

2.5.7   Linearity

Figure 2.25(b) shows the nonlinearity contributions of each transistor in this twofold noise-cancelation structure. The nonlinearity of each transistor, including its second- and third-order products, is modeled with a nonlinear current source. For a small-signal operation, the nonlinear transconductance of a short-channel CMOS transistor is represented by a power series:

$$i_m = g_m V_{gs} + \frac{g'_m}{2} V_{gs}^2 + \frac{g''_m}{6} V_{gs}^3, \tag{2.60}$$

where g_m is a small-signal transconductance, and g'_m and g''_m are its higher-order coefficients, which define the strengths of the corresponding nonlinearity. Figure 2.25(b) shows that the nonlinearities of M_1, M_7, and M_9 generate voltages at nodes X, Y, and Z, which are defined based on Volterra series as $V_{X,NL}$, $V_{Y,NL}$, and $V_{Z,NL}$, respectively. These nonlinear voltages will be canceled at the output node after passing through the three paths as shown in Fig. 2.25(b). This happens if the nonlinear coefficients a, b, and c are properly defined with Volterra series analysis, leading to gain distortion conditions related to $g'_{m_{1,7}}$ and $g''_{m_{1,7}}$.

By canceling the nonlinearity of the first stage, the linearity performance of this LNA is now only limited by the third-order distortion of M_3 and M_5. Through a long derivation, it can be proven that by properly setting the third-order nonlinearity conditions for M_3 and M_5 (e.g., via biasing), the distortion products of M_3 and M_5 will cancel each other, thus leading to a significant improvement in the total IIP3.

2.6   Conclusion

The first stage of a cellular wireless discrete-time receiver, a wideband noise-canceling low-noise transconductance amplifier (LNTA), is introduced in this chapter. In the first LNTA example, the noise-reduction technique is based on a current-reuse approach, and it is applied to the noise-cancelation stage to reduce the channel thermal noise of the following cancelation stage by increasing its transconductance, leading to improvement of the total noise figure. The second LNTA example features twofold noise-cancelation core transistor pairs: in addition to the noise of the input-matching transistor pair, the noise of another transistor pair in the LNA core is also canceled. As shown, through calculations, this LNTA could even achieve sub-1dB noise figure by increasing the gain of the LNA core and the total transconductance.

3   Discrete-Time High-Order Low-Pass Filter

One of the main building blocks in a receiver is a low-pass filter (LPF) used at the baseband. This block is responsible for selecting the desired channel. In zero-IF receivers, this block is placed directly after the radio frequency (RF) downconversion mixer. In a high-IF receiver, the LPF is required after a second downconversion from the IF to baseband. In addition to wireless communication applications [10, 11, 93, 68], integrated LPFs are the key building blocks in various other types of applications, such as hard disk drive read channels [94, 95], video signal processing [96], smoothing filtering in a DAC [97], and antialiasing filtering before a sampling system. The Noise of these filters is one of the key system-level concerns. This noise can be usually traded off with the total filter capacitance and, consequently, total power and area. Therefore, for a given system-level noise budget, a filter with a lower noise coefficient reduces the area and power consumption. On the other hand, the linearity of the filter should be high enough to maintain the fidelity of the wanted signal.

In this chapter, after a short comparison of different LPF structures, an overview of a basic DT passive LPF is presented. Next, the designed high-order DT filter [5] is described, accompanied by the design and implementation of its test chip. At the end, measurement results of the test chip are explained.

3.1   LPF Structure

As shown is Fig. 3.1, three types of commonly used analog filters are G_m-C, active RC, and active switched-capacitor (SC) filters [98–103]. In G_m-C and active RC filters, pole/zero locations are set by g_m value, capacitance (C), and resistance (R). Due to the poor matching of G_m-C and R/C values, the process–voltage–temperature (PVT) variations have a considerable impact on the filter transfer function. Therefore, many applications require component (i.e., g_m, R, and C) calibration/tuning [102, 103]. However, the pole/zero locations of active SC filters are accurately set by the capacitor ratio, thus minimizing the effect of PVT variations.

Implementation of such filters in deep nanoscale CMOS is getting increasingly difficult, especially due to the design challenges of high-quality opamps and high linearity g_m-cells. In contrast, the switching performance of MOS transistors is getting improved due to the technology scaling. Consequently, passive switched-capacitor filters are expected to work at much higher sampling rates than do the active SC

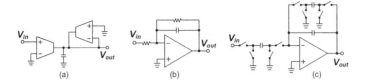

Figure 3.1 Conventional analog filters: (a) G_m-C, (b) active RC, and (c) active switched-capacitor.

filters, where the speed is limited to opamp settling. Moreover, the passive filters will consume much less power. However, it might not be possible to synthesize complex poles in a fully passive structure.

The passive LPF introduced in this chapter benefits from these advantages. By using a sampling capacitor to rotate charge between several history capacitors, a high-order IIR low-pass filter is created. To further increase the sampling rate, a pipelining technique of the sampling capacitor is introduced. Using these techniques, a seventh-order LPF is implemented, which operates up to 1 GS/s [4, 104]. In [10, 11, 92, 42, 93], passive switched-capacitor FIR/IIR filters have been used for baseband signal processing of an RF receiver. However, none of the prior publications have proposed such a high-order passive filtering in one stage. A somewhat similar structure resembling the charge-rotating filter has been reported in [105]. However, a third-order LPF filter is used in an N-path filter to form a band-pass transfer function. Furthermore, the design does not exploit any pipeline technique such as the one introduced in this work.

This filter has a very low input-referred noise, because it uses only one sampling capacitor for all the filtering stages. Thanks to the passive operation, it has an extremely high linearity. A simple inverter-based g_m-cell might be used in front of this filter to provide gain. This filter consists of only switches, capacitors, clock waveform generator, and a simple g_m-cell. Therefore, it is amenable to the digital deep nanoscale CMOS technology. The presented filter has been successfully verified at the system level in a discrete-time superheterodyne receiver [92]: The sixth-order charge-rotating filter is employed there as the first baseband channel selection filter.

3.2   Basic Discrete-Time Low-Pass IIR Filters

3.2.1   First-Order Filter

Perhaps the simplest analog discrete-time (DT) filter is a passive first-order IIR low-pass filter, as depicted in Fig. 3.2(a) [106]. In each cycle at ϕ_1, a sampling capacitor C_S samples a continuous-time input voltage $V_{in}(t)$. Hence, it is called a voltage-sampling filter. Then at ϕ_2, C_S shares its stored charge with a history capacitor C_H. At the end of ϕ_2, we have the following equation for the discrete-time output voltage:

$$V_{out}[n] = \frac{C_H}{C_H + C_S} V_{out}[n-1] + \frac{C_S}{C_H + C_S} V_{in}[n-0.5]. \qquad (3.1)$$

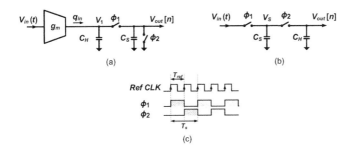

Figure 3.2 (a) Voltage-sampling and (b) charge-sampling first-order DT IIR filter with (c) their clock waveforms.

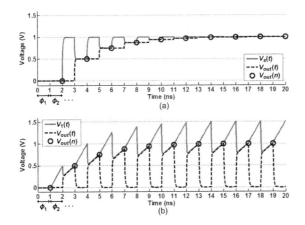

Figure 3.3 Step response of (a) the voltage-sampling and (b) charge-sampling first-order DT filter ($C_H = C_S = 1\,\text{pF}$, $g_m = 0.5\,\text{mS}$, $f_{ref} = 1\,\text{GHz}$, and $f_s = 500\,\text{MS/s}$).

Hence, its transfer function can be written in z-domain as:

$$\frac{V_{out}(z)}{V_{in}(z)} = \frac{(1 - \alpha)z^{-0.5}}{1 - \alpha z^{-1}} \tag{3.2}$$

where the coefficient α is $C_H/(C_H + C_S)$. This is a standard form of a DT low-pass filter (LPF) with unity dc gain and half-a-cycle delay, $T_s/2$. Switch driving clock waveforms are shown in Fig. 3.2(c).

The step response of this filter is shown in Fig. 3.3(a). C_S and C_H are chosen 1 pF each, just for illustration's sake. Discrete-time output samples are available in each cycle at the end of ϕ_2.

Figure 3.3(b) shows an alternative first-order DT LPF (IIR1) exploiting charge-sampling [11, 106, 107]. At first, the continuous-time input voltage is converted into current by the g_m-cell of transconductance gain g_m. This current is integrated over a time window T_s on C_H and C_S during ϕ_1 and on C_H during ϕ_2. However, we can assume for simplicity that discrete-time input charge packets arrive only at ϕ_1:

$$q_{in}[z] = \int_{(n-1)T_s}^{nT_s} g_m V_{in}(t). \tag{3.3}$$

Although this assumption slightly changes the transient waveforms of C_H and C_S voltages, it leads to exactly the same values of the output samples while simplifying the analysis of the filter. During ϕ_1, C_H shares its charge with C_S and a new charge is input. Consequently, we have the DT output samples at the end of ϕ_1:

$$V_{out}[n] = \frac{C_H}{C_H + C_S} V_{out}[n-1] + \frac{1}{C_H + C_S} q_{in}[n], \tag{3.4}$$

$$\frac{V_{out}(z)}{q_{in}(z)} = \frac{1}{C_S} \times \frac{1-a}{1-az^{-1}}. \tag{3.5}$$

The step response of this filter is drawn in Fig. 3.3(b). In this example, C_S and C_H are 1 pF, and g_m is 0.5 mS. At first, suppose that the voltage at C_H is zero. The input step voltage appears at 1 ns and causes a constant 0.5 mA current from the g_m-cell. This current is integrated on C_H during ϕ_2. Also at this time, C_S is reset to zero. Then, at ϕ_1, C_H is connected to C_S, thus sharing its charge. During ϕ_1, the input current is integrated on both capacitors. At the end of ϕ_1 (i.e., at 3 ns), an output sample of 0.5 V is produced. Likewise, it is transferred to the next cycles, thus producing 0.75 V, 0.875 V, and so on as output samples.

In the previous two structures, C_S behaves like a lossy component that leaks a time-averaged current from C_H to ground. We might call it a DT resistor (a.k.a., switched-cap resistor). This resistor in parallel with the capacitor makes a first-order low-pass filter.

Figure 3.4 shows top-level behavioral models of the IIR1 filters. In the voltage-sampling structure of Fig. 3.4(a), the sampler first samples the continuous-time (CT) analog input voltage $V_{in}(t)$ at ϕ_1 and converts it into a DT analog voltage. Then, this signal is fed to a first-order LPF with half-a-cycle delay ($z^{-1/2}$), and the output

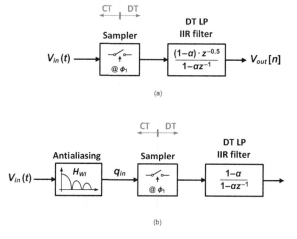

Figure 3.4 Top-level block diagram model of (a) voltage-sampling and (b) charge-sampling IIR1.

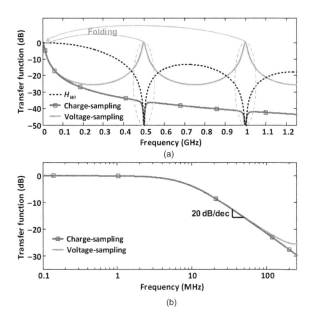

Figure 3.5 (a) Wideband transfer function and (b) Bode plot frequency response of IIR1 ($C_H = 9$ pF, $C_S = 1$ pF, $g_m = 0.5$ mS, $f_{ref} = 1$ GHz, and $f_s = 500$ MS/s).

comes out every cycle at ϕ_2. The dc voltage gain of this filter is unity. Based on the Nyquist sampling theory, the sampling of a CT signal folds frequencies around $k \times f_s$ (for $k = 1, 2, 3 \ldots$) into around dc, where f_s is the sampling frequency. As depicted in Fig. 3.5(a), we observe the folding image frequencies at f_s, $2f_s$, and so on. Figure 3.5(b) shows the transfer function, which has a roll-off of 20 dB/dec.

A behavioral model of the charge-sampling IIR1 is depicted in Fig. 3.4(b). Integrating the g_m-cell current in the time window, as described in (3.3), forms a CT sinc-type antialiasing filter prior to sampling [11, 106–108]. The transfer function of this windowed integration (WI) from the input voltage to the output charge is as follows:

$$H_{WI}(f) = g_m T_s \times \frac{\sin(\pi f T_s)}{\pi f T_s}. \tag{3.6}$$

This sinc-shape filter has notch frequencies at k/T_i ($k = 1, 2, 3, \ldots$). Assuming ideal clock waveforms, T_i is the same as $T_s = 1/f_s$. In the next step, the sampler converts the CT signal to a DT signal and, at the end, a first-order DT LPF performs the main filtering. As shown in Fig. 3.5(a), notch frequencies of the antialiasing filter are on top of the folding image frequencies, thus offering some protection. The dc voltage gain is calculated by multiplying the dc gain of the antialiasing filter by the dc gain of the DT filter:

$$A_v = \frac{V_{out}}{V_{in}} = g_m T_i \times \frac{1}{C_S f_s}. \tag{3.7}$$

In this equation, $1/(C_S f_s)$ is the equivalent DT resistance of the sampling capacitor.

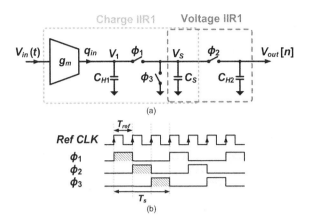

Figure 3.6 (a) Second-order DT low-pass filter and (b) its frequency ($C_H = C_S = 1\,\text{pF}, g_m = 0.33\,\text{mS}, f_{ref} = 1\,\text{GHz}$, and $f_s = 333\,\text{MS/s}$).

3.2.2 Second-Order Filter

As shown in Fig. 3.6(a), a second-order DT low-pass filter (IIR2) can be synthesized by adding a second history capacitor to the charge-sampling first-order LPF [10, 42, 68, 106]. The previously analyzed charge-sampling filter, IIR1, is indicated here within the dotted box (Charge IIR1).

At the end of ϕ_1, C_S contains the output sample of the IIR1. Then, by connecting C_S to a second history capacitor C_{H2} at ϕ_2, another first-order LPF is formed, whose structure is indicated within the dashed box (Voltage IIR1) in Fig. 3.6(a). Then at ϕ_3, the remaining history of C_S is cleared by discharging it to ground. This ensures a proper operation of the first IIR1. In this filter, the voltage-sampling IIR1 has been cascaded with the charge-sampling one, raising the total order of the filter to the second order. It should be noted that cascading two CT conventional filter stages without any loading effect would require an active buffer to isolate the first stage from the second stage. In contrast, in the DT filter of Fig. 3.6(a), there is an inherent reverse isolation between the stages through a time-slot separation, which does not require a separate active buffer. This is due to the time switching sequence and reset of C_S at the end of each cycle. In this way, the charge is only transferred from left to right, and, therefore, we thus obtain the reverse isolation. Charge-sharing equations of this filter at the end of ϕ_2 are as follows:

$$V_{out}[n] = \frac{C_{H2}}{C_{H2} + C_S} V_{out}[n-1] + \frac{C_S}{C_{H2} + C_S} V_1[n-1/3],$$

$$V_1[n-1/3] = \frac{C_{H1}}{C_{H1} + C_S} V_1[n-1/3-1] + \frac{1}{C_{H1} + C_S} q_{in}[n-1/3], \qquad (3.8)$$

which produces the following filter:

$$\frac{V_{out}}{q_{in}} = \frac{1}{C_S} \left(\frac{1 - \alpha_1}{1 - \alpha_1 z^{-1}} \times \frac{(1 - \alpha_2) z^{-1/3}}{1 - \alpha_2 z^{-1}} \right), \qquad (3.9)$$

where $\alpha_{1,2} = C_{H1,2}/(C_{H1,2} + C_S)$. This results in the same dc gain as (3.7). The transfer function of this filter is plotted in Fig. 3.6(b). The second-order IIR filter has a 2× steeper slope of 40 dB/dec compared to the IIR1 with 20 dB/dec.

3.2.3 Higher-Order Filters

Many applications require higher orders of filtering. The easiest way to build a high-order filter is to cascade two or more first- and/or second-order filters. Figure 3.7(a) shows a fourth-order filter synthesized by cascading two identical IIR2 filter stages. A similar approach has been used in [11] and [93], where two g_m-cells and passive filters are cascaded. This higher order is achieved at the cost of a higher power consumption. Since nonlinearities of the first and second IIR2 filters are added together, linearity is also worsened. Similarly, the total input-referred noise of this filter is higher than a single IIR2.

Another way of increasing the filter order is to cascade the IIR2 filter with a passive first-order switched-capacitor filter. Figure 3.7(b) shows this concept in which a third-order filter is synthesized by cascading the IIR2 and an IIR1. A similar concept is used in [11] and [91], where two passive SC filters are cascaded. The filter in Fig. 3.7(b) works as follows: At the end of ϕ_1, C_{S1} holds the sample of first-order filtered signal. Then at ϕ_2, it is connected to C_{H2} to perform charge sharing. At the same time, a second C_{S2} that was empty before is also connected to C_{H2} to resample the result of the second-order filtering. Therefore, C_S shares its charge with both C_{H2} and C_{S2}. Afterward, at ϕ_3, C_{S2}, which contains the sample of the IIR2, shares its charge with a third history capacitor C_{H3}. This sharing makes another first-order IIR filtering, which is cascaded with the previous IIR2, thus giving rise to the third-order filtering. To have proper cascading of the IIR2 and IIR1, we require a reverse isolation between them. Hence, at ϕ_1 of the next cycle, C_{S2} is discharged to zero to clear its remaining charge. This way, it does not transfer any charge back to C_{H2} at ϕ_2. The cascaded first-order

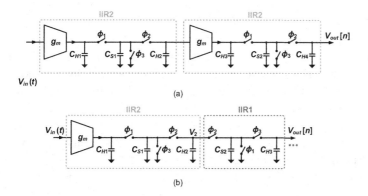

Figure 3.7 (a) Fourth-order filter synthesized by cascading two IIR2. (b) Third-order filter synthesized by resampling output of an IIR2.

filter is indicated with a dotted line at the right side of Fig. 3.7(b). Several of the IIR1 blocks can be cascaded to achieve higher orders. This filter has the following transfer function:

$$\frac{V_{out}}{q_{in}} = \frac{1}{C_{S1} + C_{S2}} \left(\frac{1 - \alpha_1}{1 - \alpha_1 z^{-1}} \times \frac{1 - \alpha_2}{1 - \alpha_2 z^{-1}} \times \frac{1 - \alpha_3}{1 - \alpha_3 z^{-1}} \times z^{-2/3} \right), \quad (3.10)$$

where

$$\alpha_1 = \frac{C_{H2}}{C_{H1} + C_{H2}}, \quad \alpha_2 = \frac{C_{H2}}{C_{H2} + C_{S1} + C_{S2}}, \quad \alpha_3 = \frac{C_{H3}}{C_{H3} + C_{S2}}. \quad (3.11)$$

The main drawback of this structure is a gain loss. Comparing (3.10) with (3.9), this third-order filter has a lower dc gain than the IIR2, caused by the second sampling capacitor C_{S2}. It leaks part of the system charge to ground in addition to the resetting of C_{S1} and, therefore, introduces more losses. The input-referred noise of this structure is also higher versus that of IIR2: first, because of the extra noise of the IIR1 part in Fig. 3.7(a); and second, because of the lowered gain of its preceding stage. In contrast, the linearity of the filter is almost the same because the switched-capacitor circuit cascaded with IIR2 is extremely linear compared to the g_m-cell active circuitry.

3.3 Novel High-Order DT IIR Low-Pass Filter

The reasoning in Section 3.2.3 makes it apparent that extending the IIR filter order using the conventional approach carries two serious disadvantages: First, the increased reset-induced charge loss lowers the gain and signal-to-noise ratio. Second, the active buffers between the stages worsen both the noise and linearity. We introduce a new structure that does not suffer from these two handicaps.

3.3.1 Charge-Rotating DT Filter

Before introducing a new high-order filter, the IIR2 is redrawn in Fig. 3.8(a). C_S is placed at the center of the (yet incomplete) circle. In each cycle, C_S is "rotating" clockwise and is sequentially connecting to C_{H1}, C_{H2}, and then the ground. To extend this idea [91], we add in Fig. 3.8(b) a few phase slots between ϕ_2 and the last reset phase, together with more history capacitors. By moving to the next new phase ϕ_3, C_S, which now holds the sample of the second-order filter, shares its charge with a third history capacitor C_{H3}. This charge sharing creates another IIR1, cascaded with the previous IIR2. Hence, we now have a third-order filtering function on C_{H3} that can be read out at the end of ϕ_3. We can continue doing so until the seventh history capacitor C_{H7} (or arbitrarily higher), in order to make a seventh-order filter. In the last phase, ϕ_8, C_S is finally reset. Since the C_S capacitor rotates the charge between the history capacitors, we call this structure a "charge-rotating" DT filter. As shown at the bottom of Fig. 3.8(c), the required multiphase clock waveforms to drive the switches can be generated from a reference clock.

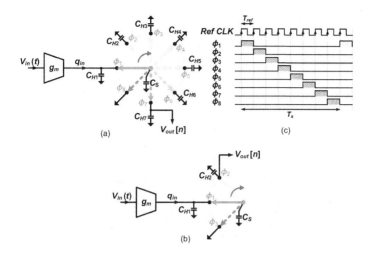

Figure 3.8 (a) The IIR2 is redrawn. (b) Charge-rotating seventh-order filter with (c) its clock waveforms. A closed switch is shown with a solid arrow, and an open switch is shown with a dimmed dashed.

Proper cascading of seven first-order IIR filters in this structure requires reverse isolation between them. This reverse isolation is provided by rotating C_S located at the center of the structure, which rotates only in one direction (i.e., clockwise here). Moreover, the resetting phase at the end of each cycle is necessary to prevent transferring charge from the last stage, C_{H7}, to the first stage at the next cycle.

Compared to the IIR2 structure in Fig. 3.6(a), the new charge-rotating (C_R) structure preserves its gain and linearity even at much higher filtering orders. The gain remains the same simply because no additional charge loss occurs in the system. In this filter structure, the switched-capacitor circuit is remarkably linear, and so the g_m-cell appears to be the bottleneck of the linearity. Moreover, the C_R filter has the same noise as IIR2, which will be discussed in Section 3.4.

3.3.2 Step Response

To better understand the operation of the filter, its step response is plotted in Fig. 3.9. At first, suppose all the capacitors are empty. For simplicity, we choose $C_S = C_H = 1$ pF. Moreover, we suppose that the input charge packet $q_{in}[n] = 1$ pC arrives every cycle at ϕ_1. A zoom-in of the step response is plotted in Fig. 3.9(a). At ϕ_1, the input charge is transferred to C_{H1} and C_S that set the 0.5 V potential on both capacitors. C_S, which contains a sample of the first-order filter at the end of ϕ_1, is then connected to C_{H2} at ϕ_2. The result is 0.25 V on both capacitors. Next, at ϕ_3, C_S, containing the sample of the second-order filter, is connected to C_{H3} and the result is 0.1 25 V. In this way, C_S transfers charge from one history capacitor to the next until C_{H7}. Then, it gets reset at ϕ_8. As plotted in Fig. 3.9(b), the outputs of higher-order stages are growing more slowly. This is because their respective input sample has been accumulated several times earlier, meaning a slower but longer and smoother integration.

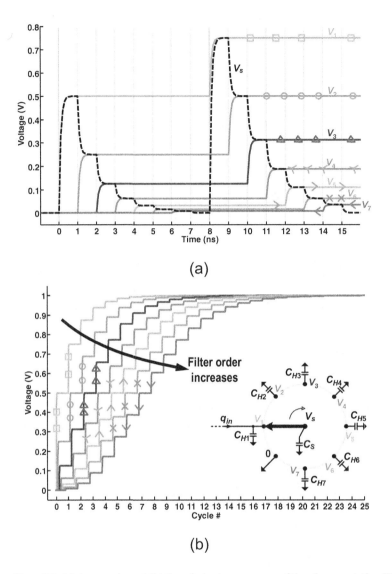

Figure 3.9 (a) A zoom-in and (b) the whole step response of the charge-rotating IIR7. ($C_H = C_S = 1\,\text{pF}$, $g_m = 0.125\,\text{mS}$, $f_{ref} = 1\,\text{GHz}$, and $f_S = 125\,\text{MS/s}$).

3.3.3   Transfer Function

Considering that samples of the main output $V_{out} = V_7$ are ready at the end of ϕ_7, we have:

$$\phi_7: V_7[n] = \frac{C_{H7}}{C_{H7} + C_S} V_7[n-1] + \frac{C_S}{C_{H7} + C_S} V_6[n-1/8]$$

$$\phi_6: V_6[n-1/8] = \frac{C_{H6}}{C_{H6} + C_S} V_6[n-1/8-1] + \frac{C_S}{C_{H6} + C_S} V_6[n-2/8]$$

$$\vdots \tag{3.12}$$

$$\phi_2: V_2[n-5/8] = \frac{C_{H2}}{C_{H2}+C_S}V_2[n-5/8-1] + \frac{C_S}{C_{H2}+C_S}V_1[n-6/8]$$

$$\phi_1: V_1[n-6/8] = \frac{C_{H1}}{C_{H1}+C_S}V_1[n-6/8-1] + \frac{1}{C_{H1}+C_S}q_{in}[n-6/8].$$

In these equations, each $-1/8$ in the discrete-time argument means one phase delay. At ϕ_7, V_7 is a function of its value at previous cycle (-1 delay) and a sample V_6 that comes from the previous phase ($-1/8$ delay). Likewise, charge-sharing equations from ϕ_1 to ϕ_6 are derived. Converting all these equations into z-domain, we can derive the following general equation for different outputs:

$$H_k(z) = \frac{V_k}{q_{in}} = \frac{1}{C_S}z^{-(k-1)/8}\prod_{i=1}^{k}\frac{1-\alpha_i}{1-\alpha_i z^{-1}}, \tag{3.13}$$

for $k = 1, 2, \ldots, 7$. In this equation, $\alpha_i = C_{H,i}/(C_{H,i}+C_S)$. Normally, we prefer to have all the poles identical, and so we choose all the history capacitors of the same size $C_{H1-7} = C_H$. Then the transfer function of the main output (i.e., V_7) is simplified to:

$$H_k(z) = \frac{V_{out}}{q_{in}} = \frac{1}{C_S}z^{-6/8}\left(\frac{1-\alpha}{1-\alpha z^{-1}}\right). \tag{3.14}$$

From this equation, the dc gain of V_{out} from the input charge, q_{in}, is $1/C_S$. Then, by using (3.7), the overall dc gain of this filter from the input voltage to its output is as follows:

$$A_v = \frac{V_{out}}{V_{in}} = g_m T_i \times \frac{1}{C_S} = g_m \times \frac{1}{C_S f_s}. \tag{3.15}$$

In this equation, $T_i = T_s$ is the time period of the cycle, that is, the eight phases. Moreover, $1/(C_S f_s)$ is an equivalent dc resistance of the sampling capacitor. This filter has the same dc gain as the IIR2 filter in (3.7).

For frequencies much lower than f_s, we can use a bilinear transform to obtain the continuous-time transfer function of the filter:

$$\frac{V_k(s)}{V_{in}(s)} = A_V \times \frac{1}{(1+\frac{1}{C_S f_s}C_H s)^k}. \tag{3.16}$$

This equation is similar to the transfer function of an RC LPF, that is, $1/(1+RC_S)$. Poles of this equation are all located at $s = -C_S f_s/C_H$. It indicates that the bandwidth of this filter only depends on the ratio of capacitors and the sampling frequency, thus making it much less sensitive to PVT variations. This salient feature eliminates any need of calibration, which is necessary for other filter types [98–103].

Wideband transfer function of this filter is plotted in Fig. 3.10(a). Similar to Fig. 3.4(b), the antialiasing filter attenuates signals around f_s, $2f_s$, and so on, before the sampling folds them to dc. Furthermore, the transfer functions at the outputs of different orders are shown in Fig. 3.10(b). The slope of the seventh-order output transfer function reaches a maximum of 140 dB/dec for far-out frequencies.

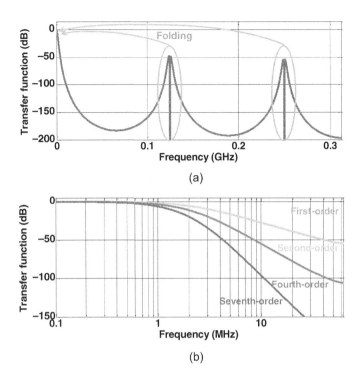

Figure 3.10 Bode plot frequency response of the C_R IIR7.
$(C_H = 9\,\text{pF}, C_S = 1\,\text{pF}, g_m = 0.125\,\text{mS}, f_{ref} = 1\,\text{GHz}, \text{and } f_s = 125\,\text{MS/s}).$

3.3.4   Equalization of the Transfer Function

In many applications, the wanted signal could be accompanied by a strong interferer. Analog-intensive receivers traditionally use continuous-time (CT) Butterworth
or Chebyshev type of filters with complex conjugate poles to select the wanted channel
out of adjacent channels while filtering out interferers and blockers. In this way,
most of the filtering is done in the CT analog domain, and a low dynamic range
ADC can be used afterward. However, digitally intensive DT receivers distribute the
channel select filtering between the pre-ADC analog filter and post-ADC digital filter.
In [10, 11, 91, 92, 42, 93, 68], second/third/sixth-order real-pole analog filters are
used before the ADC, and the rest of filtering is done in digital domain with minimum
power consumption. Considering a 3 dB BW, the transfer function of real-pole filters
exhibits a gradual and smooth transition region between the flat passband into the
sharp roll-off (see Fig. 3.11(b)). Therefore, the real-pole filters are used mostly to filter
far-out interferers/blockers, while they have a moderate selectivity between wanted
and adjacent channels.

The proposed DT C_R filter could be converted at the system level to a sharp high
selectivity filter (e.g., Butterworth) with digital assistance in the form of post-emphasis
equalizer. The idea is to "pull in" the 3 dB cutoff frequency transition region of the
analog filter to well within the channel and digitally compensate for the extra droop at

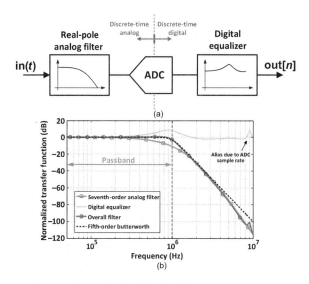

Figure 3.11 (a) Novel system for digital equalization of the filter transfer function. (b) An example of equalizing transfer function of a seventh-order real-poles DT C_R filter and comparison with a fifth-order Butterworth.

the channel edges. The gradual roll-off region of the analog filter is masked by flattening it out in the digital domain such that only the sharp roll-off remains. Figure 3.11(a) shows the concept. The digital equalizer can be an all-pass IIR filter with 0 dB gain and a small peaking at a certain frequency, thus of insignificant incremental area and power penalty, especially in scaled CMOS. The transfer function of this filter is easily calculated by dividing the targeted total transfer function by the transfer function of the analog C_R filter. In practice, its transfer function is merged with the existing digital part of the channel select filtering, sample-rate decimation, VGA, offset cancelation, I/Q mismatch compensation, and demodulation [10, 42, 68].

An example is shown in Fig. 3.11(b). The equalizer is designed to map the seventh-order real-pole transfer function of the C_R filter to a fifth-order Butterworth filter. The goal of this mapping is to flatten the passband of the overall transfer function, while keeping it unchanged or better for far-out frequencies. To maximally reduce power consumption, the digital equalizer operates in this example at a decimated rate of 10 MS/s while the analog C_R filter runs at 800 MS/s. Note that the C_R filter also serves as an effective antialiasing filter for the decimation.

The overall transfer function (including the analog filter and the digital equalizer) has a higher 3 dB bandwidth ($BW_{overall}$) than the real-pole analog filter itself (BW_{analog}). Consequently, the input signal undergoes some attenuation by the analog filter inside the overall passband, which is compensated by the small peaking of the digital equalizer. While the signal experiences an overall flat transfer function within the passband, the peaking increases noise at the transition region frequencies of both the analog filter and the ADC to some extent. To be able to compare the overall filter with a stand-alone CT analog filter, we consider that the overall filter transfer function

Table 3.1. Mapping of the seventh-order real-pole IIR filter to the overall (i.e., analog and digital) fifth- to seventh-order Butterworth filter of 1 MHz 3 dB bandwidth ($BW_{overall}$)

Order of mapped Butterworth (1 MHz BW)	Fifth	Sixth	Seventh
3 dB BW of analog filter (seventh-order, real-pole)	490 KHz	415 KHz	340 KHz
Gain loss (RMS averaged within 1 MHz BW)	4.9 dB	7.2 dB	11.2 dB
Attenuation of analog filter at 2/4 MHz	31/64 dB	37/75 dB	43/85 dB

is lumped before the ADC, but with a gain loss caused by the analog filter. This loss is equal to the RMS averaged value of the transfer function of the digital equalizer within $BW_{overall}$.

Table 3.1 summarizes the analog gain losses for three different digital equalizers that map the seventh-order real-pole IIR filter to the fifth- to seventh-order Butterworth filters with a target 3 dB $BW_{overall}$ of 1 MHz. Due to the large over-sampling ratio (800 MS/s compared to the filter BW), the reported gain losses remain almost the same in case BW_{analog} and $BW_{overall}$ are scaled proportionally. In equalization into fifth-order Butterworth, for example, a 490 kHz BW_{analog} is used to reach 1 MHz $BW_{overall}$, while 4.9 MHz BW_{analog} can be used for 10 MHz $BW_{overall}$. Since in both cases the filtering profile is similar, the passband gain loss remains almost the same 4.9 dB (from Table 3.1). Depending on the application, the order of the analog filter and the mapped transfer function should be chosen in a way that provides enough analog stopband attenuation and that minimizes the gain loss.

As this filter has passband loss for specific spot frequencies, it increases input-referred noise of the ADC at such spot frequencies. Therefore, a straightforward suggestion (i.e., rule of thumb) is to compensate this effect by increasing gain of the frond-end (before the filter) or the filter itself by the same amount. This suggestion holds very well for reasonably small droop values and when decreasing the filter order (e.g., going from seventh-order real-pole to fifth-order Butterworth).

However, in a communication channel one needs to consider interferers/blockers. There, not only the channel noise of the ADC is considered, but also its dynamic range (DR) in order to absorb the residual blockers. As shown in Fig. 3.11(b), "absolute" out-of-band blocker attenuation of the real-pole filter is the same (or better) than the mapped Butterworth filter. However, the real-pole filter has an average passband gain loss. Therefore, its "relative" blocker attenuation is lowered by this loss. Figure 3.12 compares the required ADC dynamic range for two cases of a baseband filter without and with passband loss (i.e., Butterworth and real-pole filters, respectively). When the filter has a passband loss, the dynamic range of the blockers at its output is increased by the same amount (see Fig. 3.12). Therefore, the ADC full-scale remains the same. However, the required ADC channel noise is reduced. Thus, the ADC needs an extra dynamic range equal to the amount of average passband gain loss. If one uses a higher front-end gain to return the ADC channel noise to its original value, it increases the ADC full-scale level by the same amount. Hence, in this case, the ADC requires

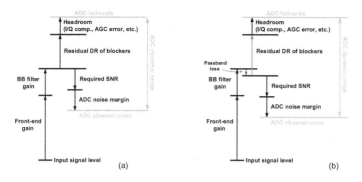

Figure 3.12 ADC dynamic range calculation considering a baseband filter (a) without, and (b) with passband loss.

the same amount of extra dynamic range. Therefore, the ADC could require 0.8–1.8 extra ENOB (1 bit per 6 dB loss) when the transfer function is mapped to the fifth- to seventh-order Butterworth, respectively. However, under certain circumstances in which the equalized TF has a higher attenuation than the desired one (e.g., far-out blockers and when decreasing the filter order from seventh-order real-pole to fifth-order Butterworth; for example, for frequencies higher than 4 MHz in Fig. 3.11(b)), the DR loss can be prevented altogether.

3.3.5   Sampling Rate Increase

Sampling rate of the C_R filter in Fig. 3.8(b) is one sample per cycle, with each cycle comprising eight phases. Therefore, the sampling frequency f_s is $f_{ref}/8$. By increasing the sampling rate, the frequency folding would be pushed higher, thus making it less of a concern. Moreover, the filter can achieve a wider bandwidth.

Operation of the C_R IIR7 filter as shown in Fig. 3.8(b) can be considered as eight different stages in series. As new data arrives at ϕ_1, it is sequentially processed at each stage until ϕ_8. Only then the next data sample arrives. In this way, we have not used the full capacity (data rate) of each stage. For example, while the data is being processed at ϕ_7 on C_{H7}, other capacitors, C_{H1} to C_{H6}, are unused awaiting a new sample. As history capacitors C_{H1-7} are holding the data between different stages, we are able to readily increase the data rate by pipelining.

Suppose that instead of only one C_S, we now have eight sampling capacitors, each of them connected to one of the "history" nodes. Then, by going to the next phase, all of them are moving to the next node in the clockwise direction. At each new phase of this pipeline structure, a new data ($q_{in}[n]$) comes into C_{H1}, a new data is transferred from C_{H1} to C_{H2}, from C_{H2} to C_{H3}, and so on until C_{H7}, and one sampling capacitor is reset to ground. Therefore, new data comes in and new data comes out at each phase (instead of each cycle). In this way, functionality of the filter has not changed while its sampling rate has increased by eight times ($f_s = f_{ref}$). Schematic of this full-rate C_R IIR7 filter is shown in Fig. 3.13. For each sampling capacitor and its rotation network, a separate switch bank is used.

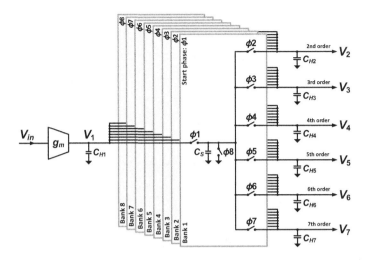

Figure 3.13 Full-rate C_R IIR7 filter using pipelining ($f_s = f_{ref}$).

The pipeline SC structure has the same charge sharing, transfer function, and gain equations as (3.8)–(3.11), but with replacing each 1/8 delay with a unit delay and considering the new f_s.

In this filter, if there is a mismatch between different C_H, it would slightly shift the pole locations. Since these capacitors have typically a large value and are of the same type, they are very well matched, thus removing the matching concern. However, if the mismatch exists between the different C_S in the pipeline structure, it could alias some amount of signal from harmonics of $f_{ref}/8$ inside the passband. However, any signal around the harmonics is filtered before the aliasing. In practice, this nonideal effect is too small to be observed.

3.3.6 Robustness to PVT Variations

Active-RC and g_m-C filters are quite sensitive to PVT variations because of poor matching between different types of elements (i.e., resistor, capacitor, and g_m-cell). However, switched-capacitor filters are quite robust to PVT variations. Transfer function and BW of SC filters are set by the capacitor ratio, which are normally implemented of the same device type (e.g., MOS, MiM, or MoM capacitor). Active-SC filters are very robust to PVT, especially when parasitic capacitance cancelation techniques (e.g., correlated double sampling) are used.

In the designed passive SC filter, the effective C_S is provided by MoM type of capacitor and also the parasitic capacitance of eight MOS switches connected to it (see Fig. 3.13). In this design, 8–26% of C_S is the MOS parasitic capacitance, depending on a C_S value selection code. On the other hand, C_H has also some switches to select its value. In this design, the MOS parasitic capacitance connected to C_H ranges from 0.5–20%, depending on the C_H selection code. Being subject to the PVT variation,

MoM part of the effective C_S and C_H tracks each other very well. In addition, their common percentage part of the remaining MOS parasitic capacitance matches well (they are of the same type). The only part that could be affected by PVT variation is the difference between the MOS parasitic capacitance percentages of C_S and C_H. Depending on the selected C_S and C_H codes, this difference is limited to a few percent of the whole capacitance. In this way, the PVT variation effect is reduced, but still somewhat higher than an active-SC filter.

3.4   Noise Analysis

Output noise of the charge-rotating seventh-order DT filter contains two main contributors: noise of the input g_m-cell and noise of the passive switched-capacitor network.

3.4.1   Noise of Transconductance Cell

Figure 3.14 shows a top-level signal flow diagram of the high-order IIR filter. At first, assume that V_{in} is zero for this noise analysis. $\overline{V_{ng_m}^2}$ is an input-referred noise of the g_m-cell (see node A in Fig. 3.14). Output noise of the g_m-cell is shaped by HWI in (3.6) before sampling (node B in Fig. 3.14). The sampling process folds frequencies higher than $f_s/2$ to the fundamental 0-to-$f_s/2$ range. Since the noise in various bands is uncorrelated, their power is added up. It can be shown that the output is a flat noise for a white input noise [91] (node C in Fig. 3.14). Power spectral density (PSD) of the sampled noise charge (q_{in}) can be found by equating power of the sampled noise with power of the shaped noise before sampling:

$$\int_0^{f_s/2} \overline{Q_{n,in}^2} df = \int_0^{inf} \overline{V_{ng_m}^2} \times \|H_{WI}(f)\|^2 df. \tag{3.17}$$

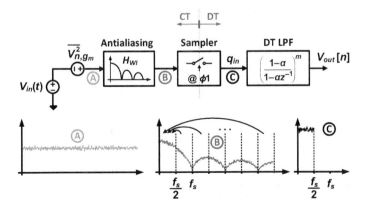

Figure 3.14 Block diagram model of the C_R IIR7 for mth-order output. (A) input-referred noise of the g_m-cell. (B) The noise shaped by antialiasing filter. (C) Sampled noise fed into DT filter.

By substituting (3.6) into (3.17) and considering $T_i = 1/f_s$ in our case, the noise PSD of the sampled input charge simplifies to.

$$\overline{Q^2_{n,in}} = \frac{g_m^2}{f_s^2} \overline{V^2_{ngm}}, \tag{3.18}$$

The above noise is fed to the switched-capacitor filter and is shaped by its transfer function:

$$\overline{V^2_{n,out}} = |H(z)|^2 \times \overline{Q^2_{n,in}}. \tag{3.19}$$

For example, the output voltage noise PSD of the C_S IIR7 can be calculated by substituting (3.14) and $z = e^{j\Omega} = e^{j2\pi f/f_s}$ into (3.19):

$$\overline{V^2_{n,out}} = \left\| \frac{1}{C_S} \cdot e^{-j\frac{6}{8}\Omega} \cdot \left(\frac{1-\alpha}{1-\alpha\,e^{-j\Omega}} \right)^7 \right\|^2 \times \frac{g_m^2}{f_s^2} \overline{V^2_{n,gm}}$$

$$= \left(\frac{1-2\alpha+\alpha^2}{1-2\alpha\cos\Omega+\alpha^2} \right)^7 \times \left(\frac{g_m}{C_S f_s} \right)^2 \overline{V^2_{n,gm}}, \tag{3.20}$$

where $g_m/(C_S f_s)$ is the voltage gain of the filter calculated also in (3.15).

3.4.2   Noise of Switched-Capacitor Network

The second key noise contributor of the filter is from the switched-capacitor network. Before calculating this noise, we first discuss the noise of a voltage-sampling process.

In Fig. 3.15(a), a voltage sampler that includes the noise of its switch is drawn. In this circuit, assume that V_{in} is zero. When the switch is on, it has a finite resistance (R_{on}). A series voltage source models the resistor's thermal noise with a constant PSD, as shown in Fig. 3.15(b):

$$S_R(f) = 4kT\,R_{on} \qquad f \geq 0, \tag{3.21}$$

where k is Boltzmann's constant and T is the absolute temperature. When the switch is on, the noise of the resistor is shaped by the RC filter with a time constant of $\tau = R_{on}C_S$ and then appears at the output. At the moment the switch is disconnected,

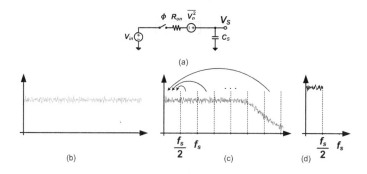

Figure 3.15 (a) Noise circuit model of a voltage-sampling process. (b) Noise of switch.

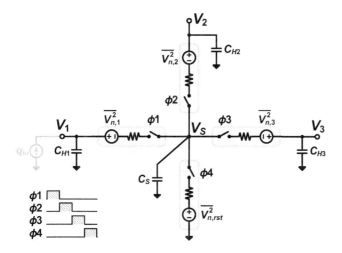

Figure 3.16 Noise model of a charge-rotating third-order filter.

the output noise is sampled and held on C_S. The sampling causes noise folding from frequencies higher than $f_s/2$, to the 0-to-$f_s/2$ range and they are added up, as shown in Fig. 3.15(c). If the time constant τ is much lower than "on" duration of the switch, it can be shown that the summation of all folded noise will be flat (i.e., white noise) [109]. As shown in Fig. 3.15(d), single-sided noise spectral density of the sampled noise at the output is [109]:

$$\overline{v_n^2}(f) = \frac{kT}{C_S f_s/2} \qquad 0 \le f \le f_s/2. \tag{3.22}$$

If we integrate this noise over the entire frequency range, its power is kT/C_S. To simplify the problem for more complicated switched-capacitor circuits, we can use the following assumption: the continuous-time noise source $\overline{V_n^2}$ with PSD of (3.21), can be considered as a discrete-time noise source with PDS described in (3.22). In this way, it is not anymore necessary to consider the effect of RC filter in switched-capacitor noise analysis.

To calculate the output noise of the C_S IIR7, we first start with a lower order for simplicity, and then extend it to the seventh order. Figure 3.16 shows a charge-rotating third-order filter. In this filter, the input current signal is considered zero. The aim of the following calculation is to find the DT output noise at different outputs, that is, V_1, V_2, and V_3, generated by noise sources $\overline{V_{n,rst}^2}$, $\overline{V_{n,1}^2}$, $\overline{V_{n,2}^2}$, and $\overline{V_{n,3}^2}$. To simplify the equations, for the rest of the calculation, we define:

$$\alpha_{1-3} = \frac{C_{H,1-3}}{C_{H,1-3} + C_S}. \tag{3.23}$$

As defined in (3.22), PSD of $\overline{V_{n,rst}^2}$ is

$$\overline{v_{n,rst}^2}(f) = \frac{kT}{C_R f_s/2}. \tag{3.24}$$

However, other noise sources are differently calculated. Since from ϕ_1 to ϕ_3, C_S is in series with C_{H1} to C_{H3}, the total capacitance at each phase should be taken into account for noise PSD:

$$\overline{v_{n,1-3}^2}(f) = \frac{kT}{\frac{C_{H,1-2}C_S}{C_{H,1-3}+C_S}f_s/2} = \frac{kT}{\alpha_1 - 3\,C_S f_s/2} = \frac{\overline{v_{n,rst}^2}}{\alpha_{1-3}}. \tag{3.25}$$

Sampling frequency, f_s, in (3.24) and (3.25) is the repetition frequency of each phase, equal to $f_{ref}/4$ in this case. At ϕ_4, C_S is reset. In other words, the effects of noise sources at other phases on C_S are all cleared. At the end of this phase, it samples noise of the reset switch $\overline{V_{n,rst}^2}$:

$$@\phi_4 : v_s[n] = v_{n,rst}[n]. \tag{3.26}$$

Then, C_S is connected to C_{H1} at ϕ_1. Charge-sharing equations at this phase are:

$$@\phi_1 : v_1[n] = \frac{C_{H1}}{C_{H1} + C_S}v_1[n-1] + \frac{C_S}{C_{H1} + C_S}v_{n,1}[n] + \frac{C_{H1}}{C_{H1} + C_S}v_s[n-1/4], \tag{3.27}$$

where $v_1[n-1]$ is the previous history of C_{H1}, $v_S[n-1/4]$ is a voltage sample of V_S from the previous phase (i.e., α_4), and $v_{n,1}[n]$ is noise of switch at α_1. Combining (3.26) and (3.27) and (3.23), we obtain:

$$@\phi_1 : v_1[n] = \alpha_1 v_1[n-1] + (1-\alpha_1)v_1[n-1]v_{n,1}[n]$$
$$+ (1-\alpha_1)v_{n,rst}[n-1][n-1/4]. \tag{3.28}$$

Now, we can calculate the noise transfer function to output V_1 by using a z-transform:

$$V_1 = \frac{1-\alpha_1}{1-\alpha_1 z^{-1}}V_{n,1} + \frac{(1-\alpha_1)z^{-\frac{1}{4}}}{1-\alpha_1 Z^{-1}}V_{n,rst}. \tag{3.29}$$

To see PSD of V_1, we substitute $z = e^{j\Omega}$ and it follows that:

$$\overline{V_1^2}(e^{j\Omega}) = \left\| \frac{1-\alpha_1}{1-\alpha_1 e^{-j\Omega}} \right\|^2 \overline{V_1^2} + \left\| \frac{(1-\alpha_1)e^{-\frac{1}{4}j\Omega}}{1-\alpha_1 e^{-j\Omega}} \right\|^2 \overline{V_{n,rst}^2}. \tag{3.30}$$

Then, it reduces to:

$$\overline{V_1^2}(e^{j\Omega}) = \frac{1 - 2\alpha_1 + \alpha_1^2}{1 - 2\alpha_1 \cos\Omega + \alpha_1^2} \times (\overline{V_{n1}^2} + \overline{V_{n,rst}^2}). \tag{3.31}$$

Before going to the next phase, we need to calculate the remaining noise on C_S at the time it is disconnecting from C_{H1}:

$$@\phi_1 : v_s[n] = v_1[n] - v_{n,1}[n]. \tag{3.32}$$

By using (3.29), its noise transfer function on V_S at the end of ϕ_1 is

$$@\phi_1 : V_S = -\frac{\alpha_1(1 - z^{-1})}{1 - \alpha_1 z^{-1}} V_{n,1} + \frac{(1 - \alpha_1)z^{-\frac{1}{4}}}{1 - \alpha_1 z^{-1}} V_{n,rst}. \tag{3.33}$$

Then, its noise PSD is simplified to

$$\overline{V_S^2}(e^{j\Omega}) = \frac{\alpha_1(2 - 2\cos\Omega)}{1 - 2\alpha_1\cos\Omega + \alpha_1^2} \left(\overline{V_{n1}^2} + \frac{1 - 2\alpha_1 + \alpha_1^2}{1 - 2\alpha_1\cos\Omega + \alpha_1^2} \overline{V_{n,rst}^2} \right). \tag{3.34}$$

Substituting (3.25) in this equation, it reduces to:

$$\overline{V_S^2}(e^{j\Omega}) = \left(\frac{\alpha_1(2 - 2\cos\Omega)}{1 - 2\alpha_1\cos\Omega + \alpha_1^2} \times \frac{1}{\alpha_1} + \frac{1 - 2\alpha_1 + \alpha_1^2}{1 - 2\alpha_1\cos\Omega + \alpha_1^2} \right) \times \left(\overline{V_{n,rst}^2} \right)$$

$$= 1 \times \overline{V_{n,rst}^2} \tag{3.35}$$

This appears to be a very interesting and important result. It suggests that the noise PSD of V_S at ϕ_1 is exactly the same as its PSD at previous phase, ϕ_4, which is the reset phase. It can be explained as follows: at the end of ϕ_4, V_S has a PSD of $\overline{V_{n,rst}^2}$, which is a flat noise. Next, when C_S is connected to C_{H1} at ϕ_1, this noise is low-pass filtered (the second term of (3.34)). However, at this phase, a high-pass filtered noise originated from $\overline{V_{n,1}^2}$ is added to V_S (the first term of (3.34)). These two noise contributions are shown in Fig. 3.17(a). The latter noise compensates for the attenuated part of the former noise in a way that the total PSD remains constant and is equal to the original one, $\overline{V_{n,rst}^2}$. This result is independent of the history capacitor value.

Similar to what was described for ϕ_1, the same set of equations, (3.27) and (3.35), are valid for other phases executing before the reset phase (ϕ_4 in this case). Therefore, in general we have:

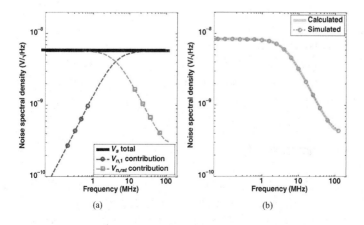

(a)   (b)

Figure 3.17 (a) Calculated noise spectral density of V_S at the end of ϕ_1 and (b) noise spectral density of V_1 ($C_H = 9\,\text{pF}$, $C_S = 1\,\text{pF}$, $g_m = 0.125\,\text{mS}$, $f_{ref} = 1\,\text{GHz}$, and $f_s = 125\,\text{MS/s}$).

$$\overline{V_i^2}(e^{j\Omega}) = \frac{1 - 2\alpha_i + \alpha_i^2}{1 - 2\alpha_i \cos\Omega + \alpha_i^2} \times (\overline{V_{n,i}^2} + \overline{V_{n,rst}^2}), \tag{3.36}$$

for $i = 1, 2, 3$ in the C_R third-order filter. Moreover, V_S at the end of each phase has the same noise PSD as calculated in (3.35).

Although it would seem at first that the noise of higher-order outputs should be increased due to the accumulation of noise coming from different noise sources, surprisingly, (3.36) rejects this hypothesis. Suppose all the history capacitors are equal, then noise PSD of all different outputs becomes the same. In other words, it does not build up by going to higher orders. This advantage of DT C_R filter is in contrast with conventional filters. For example, in an active-RC filter, more resistors and opamps are required to increase an order, thus leading to a higher output noise. Figure 3.17(b) plots noise spectral density of different outputs with equal C_H. To validate the previous equations, noise of the switched-capacitor circuit is simulated by means of Cadence SpectreRF™ PNOISE simulation. Simulation results match very close to (3.36). An interesting point is that spot noise of the filter for frequencies inside the BW is

$$\overline{V_i^2}|_{in-band} = 4kT \times \frac{1}{C_S f_s} \tag{3.37}$$

for α near 1, which is noise of the equivalent resistor, $R_{eq} = 1/(C_S f_s)$. Another interesting point is about the total noise power of each output:

$$P_{n,i} = \int_0^{f_s/2} \overline{V_i^2}\left(e^{j2\pi f/f_s}\right) df = \int_0^{f_s/2} \frac{1 - 2\alpha_i + \alpha_i^2}{1 - 2\alpha_i \cos\Omega + \alpha_i^2} \times \left(\overline{V_{n,i}^2} + \overline{V_{n,rst}^2}\right) df$$

$$= \frac{f_s}{2} \times \frac{1 - \alpha_i}{1 + \alpha_i} \times \left(\overline{V_{n,i}^2} + \overline{V_{n,rst}^2}\right). \tag{3.38}$$

By using (3.25), it reduces to:

$$P_{n,i} = \frac{f_s}{2} \times \frac{1 - \alpha_i}{1 + \alpha_i} \times \left(\frac{1}{\alpha_i} + 1\right) \times \frac{kT}{C_S f_s/2} = \frac{kT}{C_{H,i}}. \tag{3.39}$$

This result is same as the well-known output noise power of an RC filter, that is, kT/C. All the above results and equations are valid and extendable to higher order filters, for example, the C_S IIR7 discussed before and beyond. Note that if a pipeline technique is used, all the previous equations remain the same except that the new sampling frequency should be used.

3.5   Design and Implementation

The high-order charge-rotating DT filter consists of a g_m-cell, switches, capacitors, and a clock waveform generator circuit. Therefore, it is amenable to the digital deep nanoscale CMOS technology. If we implement this filter in a finer process, area of the capacitors, switches, and the waveform generator reduces while preserving or improving the performance according to Moore's law of scaling.

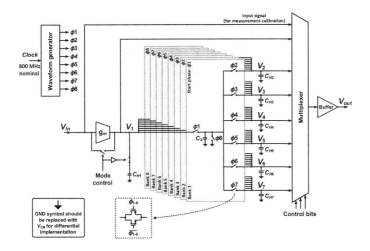

Figure 3.18 Implementation of the full-rate C_R IIR7. The circuit has been implemented differentially while it has been shown single-ended here for simplicity.

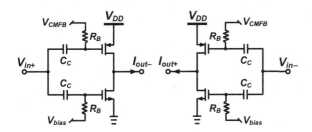

Figure 3.19 Inverter-based pseudo-differential g_m-cell.

3.5.1 Design of the Seventh-Order Charge-Rotating DT Filter

The final design is implemented differentially, while, for the sake of simplicity, it is shown single-ended in Fig. 3.18. The designed filter is software controlled to operate in one of the two modes: (1) charge-sampling, and (2) voltage-sampling. Although in the charge-sampling mode we have an active g_m-cell, the filtering network is fully passive, making the overall filter semi-passive. In the voltage-sampling mode, the g_m-cell is bypassed and disconnected from the power supply, resulting in a fully passive filter. Moreover, C_{H1} is disconnected via "mode control" to prevent loading the input. The removal of C_{H1} lowers the filtering order by one to sixth. In this mode, the input voltage (instead of the input charge) is directly sampled by C_S capacitors.

The simple inverter-based g_m-cell (Fig. 3.19) makes the filter amenable to process scaling. In this pseudo-differential g_m-cell, a bias voltage V_{bias} comes from a diode-connected NMOS and mirrors a bias current into the g_m-cell. Moreover, a feedback circuit sets the common-mode output voltage to $V_{DD}/2$ by adjusting VCMFB. Coupling capacitors C_C and bias resistors R_B set a lower limit in frequency response. By using large C_C and R_B, this limit is pulled down to a few kHz, which is acceptable for

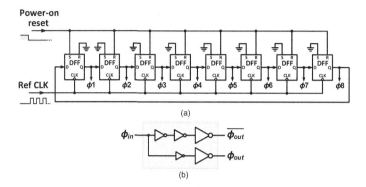

Figure 3.20 (a) Waveform generator circuit with (b) its output buffer.

most applications. As the g_m-cell, a simple inverter is used to be amenable to scaling and provide good linearity. By properly sizing of NMOS and PMOS transistors, their nonlinearities could be canceled out perfectly for square-law transistors [110]. However, in nanoscale CMOS a partial cancelation is carried out. We have used transistors with a large channel length to make their behavior closer to the square-law model. Moreover, a low resistance load by the SC circuit allows a high IIP3.

The differential history capacitors C_{H1-7} range from 0.25 to 64 pF digitally selectable via 8 bits. For both history and sampling capacitors, we used MOM capacitors to have a very good matching. This minimizes the variations due to PVT. Differential values of the sampling capacitors C_S range from 0.4 to 2.2 pF digitally selectable by 4 bits. Instead of implementing C_S differentially, we implement each as two single-ended capacitors. Then we can set common-mode voltage of the filter by terminating C_S to $V_C M$ instead of ground. To adjust the filter bandwidth, we keep C_S fixed and change C_H. In this way, both the gain and linearity of the circuit do not change. Moreover, if the sampling frequency is changed, we change C_S inversely to keep the bandwidth and the gain constant. As shown in Fig. 3.18, the filter's switches are implemented with transmission gates. We have chosen equal NMOS and PMOS sizes to reduce the charge injection and cancel out the clock feedthrough by at least an order of magnitude [108, 109], and at the same time having a lower on-resistance (R_{on}). To have a low R_{on}, low-V_T transistors are chosen. These transistors should be sized carefully to have a low enough R_{on} for fast settling on the sampling capacitors.

The waveform generator shown in Fig. 3.20(a) is a digital logic block. It consists of eight D flip-flops (DFF). At power-on, the DFFs are set/reset to "10000000". Then, at each successive clock cycle, the code is rotated one step. In this way, all the required phases are generated from a reference clock. Outputs of the DFFs are fed to buffer cells before driving the switches (Fig. 3.20(b)). The buffer is able to drive the switches with sharp rising and falling edges. Sizes of NMOS and PMOS transistors in the buffer are skewed to ensure nonoverlapping between consecutive phases.

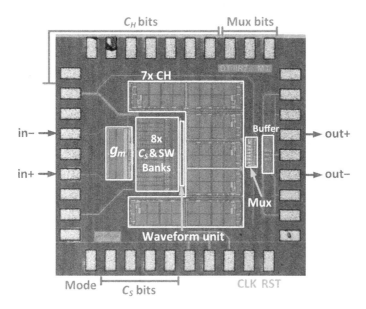

Figure 3.21 Chip micrograph of the full-rate C_R IIR7 implemented in TSMC 65 nm CMOS. Die size is 1.2×1.27 mm.

3.5.2 Implementation

The filter has been implemented in TSMC 1P7M 65 nm CMOS process. It operates at a 1.2 V power supply. The g_m-cell drains 250 μA. The waveform generator unit and its buffers clocked at a reference frequency f_{ref} of 800 MHz consume 1.40 mA. The latter current consumption is proportional to f_{ref}. The filter has been verified to work properly up to 1 GS/s. As shown in Fig. 3.18, an analog multiplexer is added to allow monitoring different outputs (orders) of the filter as well as the internal on-chip input of the filter. After the multiplexer, an output buffer isolates the chip internals from the outside. Figure 3.21 is a chip micrograph of the implemented filter.

3.6 Measurement Results

To verify the designed filter, a single-ended input signal is converted to differential with a wideband transformer, terminated with a 50 Ω resistor on the PCB and then fed to the chip. Differential output of the chip with zero-order-hold is converted back to a single-ended signal with another transformer. Table 3.1 summarizes the filter performance in its two operational modes, including the effects of the suggested digital equalizer, and compares with the recent state-of-the-art filters.

Transfer function of the filter has been evaluated using HP8753E network analyzer. To lower the measurement noise floor, a wideband RF amplifier (HP8347A)

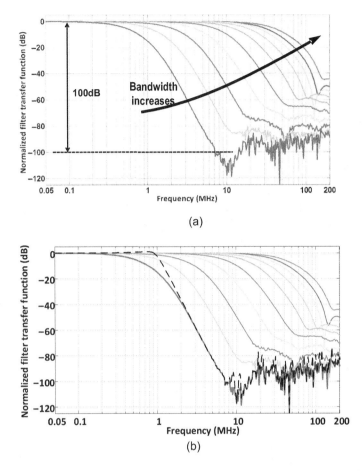

(a)

(b)

Figure 3.22 Measured transfer function of the C_R IIR7 for (a) the seventh-order output with different BW settings, and (b) for different orders in 400 kHz BW setting. The filter is in charge-sampling mode clocked at 800 MHz.

is used. Figure 3.22(a) plots the measured frequency response of the filter in the charge-sampling mode at the seventh-order output for different bandwidth settings. The 3 dB bandwidth is programmable from 400 kHz to 30 MHz. By applying a digital equalizer to map the transfer function to a fifth-order Butterworth, the overall 3 dB bandwidth would be tunable from 0.82 to 61 MHz. As an example, the overall transfer function including the equalizer is plotted with the black dashed line in Fig. 3.22(a). A maximum 100 dB of stop-band rejection is measured for the narrowest bandwidth setting. Depicted in Fig. 3.22(b) is the measured transfer function of the filter in the charge-sampling mode, but now for different outputs (orders). In this measurement, the 400 kHz analog bandwidth setting is used. The measured seventh-order output is also compared with the ideal mathematical transfer function, shown in the black dashed line, indicating a very good agreement with theory. Transfer function of the filter in the voltage-sampling mode is similar to Fig. 3.22 except that the filtering order is sixth.

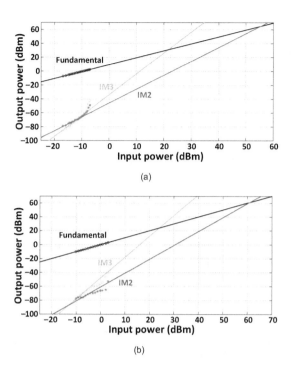

Figure 3.23 In-band IIP2 and IIP3 measurement of the C_R IIR7 filter in (a) charge-sampling and (b) voltage-sampling modes.

To evaluate linearity of the filter, a two-tone signal is fed to the filter and its output is evaluated by Agilent E4446A spectrum analyzer. For this test, the analog bandwidth is set to about 9 MHz. Figure 3.23 shows the measured second- and third-order inter-modulation products versus the input power for both charge-sampling and voltage-sampling modes. Measured in-band IIP2 and IIP3 (with respect to 50 Ω) are +55 dBm and +21 dBm in charge-sampling, and +61 dBm and +24 dBm in voltage-sampling mode, respectively. In the charge-sampling mode, where the linearity is limited by the g_m-cell, IIP3 might be lowered by a few dBs in practice, caused by PVT variations. As listed in Table 3.2, IIP3/IIP2 in the charge-sampling mode is among the best. Thanks to the fully passive operation, the filter in the voltage-sampling mode has an exceptionally high IIP3/IIP2.

Measured out-of-band IIP2 and IIP3 are +60 dBm and +11.7 dBm in charge-sampling, and +68 dBm and +25 dBm in voltage-sampling mode. IIP3 of an inverter-based g_m-cell depends also on its output resistance [50]. The g_m-cell is designed to have the best IIP3 when it is loaded with the nominal resistance seen from the switched capacitor filter. When out-of-band tones are applied, the g_m-cell sees a lower impedance than the nominal value for those frequencies, and therefore its IIP3 is reduced. This is the reason why the charge-sampling filter has a lower out-of-band linearity than in-band. However, this effect does not exist for the voltage-sampling mode, resulting in almost the same in-band and out-of-band IIP3.

Table 3.2. Performance summary and comparison with state-of-the-art

	This work [5] DT IIR @ 800 MS/s				JSSC'11 [101]	JSSC'10 [100]	JSSC'09 [99]	JSSC'06 [98]	JSSC'06 [102]	JSSC'07 [103]
	Charge-sampling		Voltage-sampling							
Topology	Without equalizer	With equalizer	Without equalizer	With equalizer	G_m-C Low power Hi. linearity	G_m-C Noise reduction	G_m-C Wide tuning	Source follower	Active RC	Active RC
Technology	65 nm		65 nm		90 nm	90 nm	0.18 μm	0.18 μm	0.12 μm	0.13 μm
Order	Seventh	Fifth	Sixth	Fourth	Sixth	Fourth	Third	Fourth	Third and fifth	Fifth
Type	Real poles	Butter.	Ral poles	Butter.	Butter.	Butter.	Butter.	Butter.	Cheb. Elliptic	Chebyshev
Supply voltage (V)	1.2		1.2		1.0	2.5	1.2	1.8	1.0	1.5
Power (mW)	1.98	–	1.68	–	4.35	1.25	4.1~11.1[a]	4.1	4.6	11.25
3 dB BW (MHz)	0.4~30	0.82~61	0.43~32	0.77~57	8.1~13.5	2.8	0.5~20	6~14	5, 10	8.9, 19.7
Gain (dB)	9.3	4.4	0	-3.5	-2.7	15	0	-3.5	0	2
Max. stopband rejection (dB)	>100		>85		90	55	60	–	50	–
In-/out-of-band IIP3 (dBm)	21.7[b]/11.7[c]		24[b]/25[c]		21.7~22.1 17.5~18.9	10/35.6	22.3~19 17.5~13[a]	8/12	18.8~21.3/-	18.3/-
In-/out-of-band IIP2 (dBm)	55[b]/60[d]		61[b]/68[d]		-51.9~-53.4	-/-	40~30.8[a] –	-/-	-/-	-/-
$P_{1dB,out}$ (dBm)	+10		+13		+3.9	–	–	0.5	-4.1~-2.6	+7
IRN (nV/$\sqrt{Hz}$)	4.57	4.72	4.97	7.27	75	22.6	425~12[a]	7.5	85, 143	30
SFDR (dB)	75~67[a]	74~66[a]	82~69[a]	77~66[a]	54.3	58	51~61	65	–	55.2
1% HD3 DR (dB)	93~81[a]	91~79[a]	103~83[a]	96~78[a]	63	67	80	79	–	69
Active area (mm²)	0.42		0.35		0.239	0.5	0.23	0.26	0.17	0.2

[a] In the same order of the filter BW.   [b] 3 MHz and 4 MHz tones.   [c] 20 MHz and 35 MHz tones.   [d] 30 MHz and 35 MHz tones.

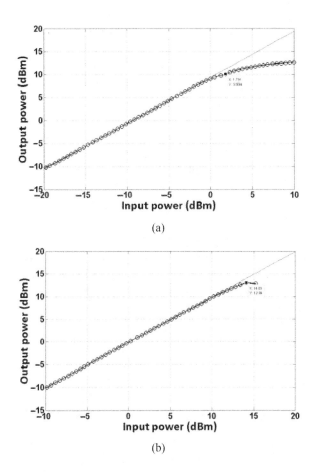

Figure 3.24 Measured P_{1dB} (a) in charge-sampling and (b) voltage-sampling modes.

Comparing in-band IIP3 of charge- and voltage-sampling modes, it seems that IIP3 of voltage-sampling should be higher than what reported (due to lower gain than charge sampling). However, the lower IIP3 than expected is caused by the extra switch that shorts the g_m-cell in voltage-sampling mode (Fig. 3.18). This switch, in series with the voltage-sampling filter, degrades the original linearity of the filter. To be able to fairly compare the 1 dB compression point of our filter in its two operational modes to other filters with various gains, we compare the output compression point $P_{1dB,out} = P_{1dB,in} + \text{Gain} - 1$. The measured output compression point of the filter in the charge-sampling mode is $+10$ dBm, while in the voltage-sampling mode, this value goes to $+13$ dBm (Fig. 3.24). These are outstanding values compared to other works listed in Table 3.2.

In charge-sampling mode, P_{1dB} of the g_m-cell is limited by its output saturation between near 0 and V_{DD}. The $P_{1dB,out}$ of $+10$ dBm is translated to 1 V peak-to-peak swing for each of the g_m-cells, while V_{DD} is 1.2 V. Therefore, in this case, P_{1dB} is not only set by third-order nonlinearity, but also with higher orders caused by output clipping of g_m-cell between ~ 0.1 V and ~ 1.1 V. In voltage-sampling mode, $P_{1dB,out}$ of $+13$ dBm is equal to ~ 1.4 V peak-to-peak single-ended swing that barely exceeds the supply limit. This happens, thanks to the fully passive structure made of only switches and capacitors. After this point, bulk diodes of the transistors are starting leaking current and compressing signal. Furthermore, $P_{1dB,in} = +14$ dBm in this mode is also very close to typical prediction of $P_{1dB,in} = $ IIP3-10.

Noise of the filter is evaluated by the spectrum analyzer. For this measurement, input of the filter is grounded. A two-step experiment is carried out: (1) measuring the total output noise (including noise of the filter and output buffer), (2) disabling the filter and measuring the noise of only the output buffer. Then, since the noise of the buffer and the filter are uncorrelated, the filter noise is calculated by subtracting the total noise PSD and the buffer noise PSD.

Figure 3.25(a) shows the measured input-referred noise (IRN) spectral density of the filter in the charge-sampling mode for the 9 MHz BW$_{analog}$ setting ($C_S = 0.75$ pF, $C_H = 2.35$ pF). The slope below 1 MHz is due to the flicker and bias noise of the g_m-cell. The noise between 1 MHz and 20 MHz is mainly the thermal noise of the g_m-cell shaped by the filter transfer function, and the rest is dominantly the noise of the switched-capacitor circuit. The averaged spot noise over the bandwidth is 4.57 nV/$\sqrt{\text{Hz}}$. Integrated noise, PN, from 50 kHz to 9 MHz is $13.6\,\mu$Vrms, which increases to $16.4\,\mu$Vrms for the entire frequency range. This gives a 71 dB

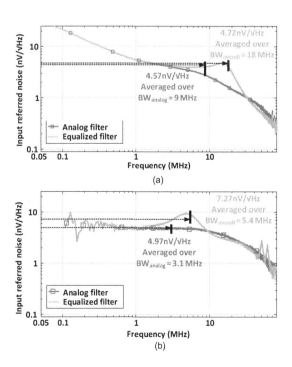

Figure 3.25 Measured input-referred noise of the filter in (a) charge-sampling and (b) voltage-sampling mode.

spurious-free dynamic range (SFDR) as defined in [71]. As measured by a single-tone test, a -3.5 dBm input signal (422 mV peak-to-peak differential) creates -40 dB third-harmonic distortion (HD3) at the output, giving an 81 dB dynamic range (1% HD3 DR). By applying the digital equalizer to map the analog transfer function to a fifth-order Butterworth, the equivalent IRN of the total filter increases to 4.72 nV/$\sqrt{\text{Hz}}$, resulting in 68 dB SFDR for 18 MHz BW$_{\text{overall}}$.

Measured input-referred noise of the filter in the voltage-sampling mode for the 3.1 MHz BW$_{\text{analog}}$ ($C_S = 0.75$ pF, $C_H = 8.25$ pF) is illustrated in Fig. 3.25(b). In this mode, the whole noise spectrum is due to the switched-capacitor network. The spot IRN averaged over the bandwidth is 4.97 nV/$\sqrt{\text{Hz}}$. The integrated IRN over the bandwidth is 8.6 μVrms and rises to 22.2 μVrms for the entire frequency range. This results in 75 dB SFDR. As measured, a single-tone input signal as large as 8.8 dBm (1.75 V peak-to-peak differential) creates 1% HD3 in this mode. This results in 97 dB dynamic range. By mapping the sixth-order real-pole transfer function of this filter to a fourth-order Butterworth using the equalizer, the overall IRN of the filter rises to 7.27 nV/$\sqrt{\text{Hz}}$ giving 71 dB SFDR for 5.4 MHz BW$_{\text{overall}}$.

Based on a hand calculation in (3.37), in-band spot noise is $4kT/(C_S f_s)$ for the switch capacitor circuit itself. In the voltage-sampling mode, this value is about 5.2 nV/$\sqrt{\text{Hz}}$, for the used $C_S = 750 \, fF$ in the filter. As shown in Fig. 3.25(b), the measured in-band spot noise is about 5 nV/$\sqrt{\text{Hz}}$, in good agreement with the hand-calculated value. Using a g_m-cell before the SC circuit with a gain of 9.3 dB, reduces the input referred noise of the SC circuit in charge-sampling mode. However, the g_m-cell itself adds some extra noise, increasing the total input-referred noise. Considering input-referred noise of g_m-cell as $4kT\gamma/g_m$, we have

$$IRN = \sqrt{\frac{4kT}{C_S f_s} \times \frac{1}{A_V^2} + \frac{4kT\gamma}{g_m}} = 3.6 \text{ nV}/\sqrt{\text{Hz}} = \frac{kT}{C_{H,i}}, \tag{3.40}$$

where $A_V = 9.3$ dB, $g_m = 1.7$ mS, and $\gamma = 1$. This is the expected value for an ideal g_m-cell. However, RC bias used in the g_m-cell further increases the noise at low frequencies (see Fig. 3.25(a)). As measured, the average spot noise in the charge-sampling mode is 4.57 nV/$\sqrt{\text{Hz}}$, aligned with the above hand calculation.

Figure 3.25 shows the measured input-referred noise in both charge-sampling and voltage-sampling modes. Measured clock feedthrough at the output of this filter is less than -110 dBm at $f_{ref}/8 = 100$ MHz. This very low value avoids any noise and spur problems caused by the clock signal.

3.7 Conclusion

In this chapter, a high-order discrete-time (DT) charge-rotating (C_R) IIR filter structure is described. It is a necessary building block of a DT superheterodyne receiver. Implementation of the seventh-order charge-sampling sixth-order voltage-sampling DT filter is elaborated in detail. The order of this filter is easily extendable to even higher orders. Using a pipelining technique, the sample rate of the filter is increased up to 1 GS/s. The C_R filter is process scalable according to the Moore's law and friendly to digital nanoscale CMOS technology. The transfer function of this filter is precise and robust to PVT variations. Even though the C_R filter features real poles, modern system applications, such as wireless receivers, could expend digital post-processing to equalize the droop at the passband edge of the transfer function. Its state-of-the-art performance includes very low power consumption, the lowest input-referred noise, very wide tuning range, and excellent linearity.

4   Discrete-Time Band-Pass Filter

A complex quadrature charge-sharing (CS) technique is utilized to implement a discrete-time band-pass filter with a programmable bandwidth of 20–100 MHz. The BPF is a natural part of a cellular superheterodyne receiver and completely determines the receiver frequency selectivity. It operates at the full sampling rate (4×) (described in Chapter 2) of up to 5.2 GHz corresponding to the 1.2 GHz RF input frequency, thus making it free from any aliasing or replicas in its transfer function. Furthermore, the advantages of CS-BPF over other band-pass filters such as N-path, active-RC, G_m-C, and biquad are described. A mathematical noise analysis of the CS-BPF and the comparison of simulations and calculations are presented. The entire 65 nm CMOS receiver, which does not include a front-end LNTA for test reasons, achieves a total gain of 35 dB, IRN of 1.5 nV/$\sqrt{\text{Hz}}$, out-of-band IIP3 of $+10$ dBm. It consumes 24 mA at 1.2 V power supply.

4.1   Introduction

Monolithic RF receivers (RX) have conventionally used a zero/low intermediate frequency (IF) due to straightforward silicon integration of low-pass channel-select filtering and avoidance of images (when zero-IF) or their easy baseband filtering (when low-IF) [22, 111, 48, 17, 112, 113]. However, their drawbacks (such as poor second-order nonlinearity, sensitivity to $1/f$ (flicker) noise, and time-variant dc offsets) are all getting ever more severe with CMOS scaling. These problems could be solved by increasing the IF frequency, as was the norm in the pre-IC era with superheterodyne radios. However, to avoid interferers and blockers at IF images, a high quality (Q)-factor band-pass filtering (BPF) is required, which is extremely difficult to implement in CMOS using continuous-time circuitry.

The integration problem of high-IF BPF was solved in [8, 6, 114]. A high-Q complex frequency translation ("N-path") filtering at the high-IF stage was used in [8] as an alternative to the conventional CT BPF. However, that filter cannot reject images defined as interferers at odd harmonics of the IF frequency because the N-path filter *inherently* features replicas there. Therefore, there is an increased demand for highly integrated BPFs that would be free from any of those replicas and still compatible with CMOS scaling suitable for superheterodyne RX. In [6, 114], we have introduced a full-rate charge-sharing (CS) discrete-time (DT) operation that is largely free from replicas and which additionally offers a freedom to change the IF frequency in face of large blockers, thus avoiding desensitization.

In this chapter, we describe in detail such high-IF DT BPF filter capable of realizing a fully integrated superheterodyne RX. The filter exploits passive switched-capacitor techniques and, as such, is amenable to CMOS scaling and is very robust to mismatches. Its center frequency and bandwidth are well controlled via clock frequency and capacitor ratios. Section 4.2 gives an overview of various types of band-pass filtering. Section 4.3 begins with the basic principles of CS-BPF and then continues with the detailed structure and continuous-time model of CS-BPF. The noise analysis of CS-BPF and circuit implementation of the front-end RX are presented in Sections 4.4 and 4.5, respectively. The measurement results are demonstrated in Section 4.6.

4.2   Overview of Band-Pass Filtering

As an overview, the transfer functions of different types of BPFs are compared in Fig. 4.1. CT filters, such as G_m-C and biquad, do not exhibit any aliasing or replicas, but their structure is very complex and they consume a lot of power. Furthermore, their input-referred noise and linearity are much worse compared to other filters due to the number of active g_m-cells used. Active-RC filters are divided into two subcategories: sample-based and continuous-time. Both use opamps or g_m-cells as active components. They typically consume a lot of power and they also tend to be large in order to reduce flicker noise generated by the active devices.

Key advantage of the full-rate CS-BPF compared to the N-path filter [115, 116, 111, 117, 118] is that its transfer function has only one peak in the entire sampling

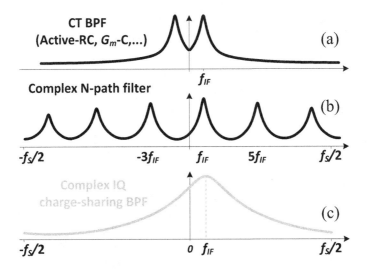

Figure 4.1 Transfer function comparison of different types of BPFs (a) CT BPF, (b) Complex N-path, and (c) DT CS-BPF.

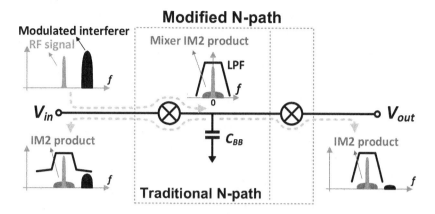

Figure 4.2 N-path filter and its second-order nonlinearity.

frequency domain of $-f_s/2$ to $f_s/2$, as shown in Fig. 4.1(c). Another advantage is that it features a theoretically infinite IIP2 compared to the limited IIP2 of N-path filters. The only drawback of DT CS-BPF compared to N-path filter is that it has a smaller Q-factor, which can be solved by cascading several CS-BPF stages or using a positive feedback [118].

The simplified block diagram of N-path filter is shown in Fig. 4.2, which comprises one mixer and baseband capacitor (C_{BB}) for a traditional N-path filter [117], or two mixers and C_{BB} for a modified N-path filter [119]. The input signal is downconverted to dc by the mixer, filtered by a low-pass filter, and then upconverted by the same [116] or another mixer [119, 118]. The second-order nonlinearity of the mixer depends on LO frequency, and any mismatch in the mixer switching transistors [120]. The typical IIP2 of the mixer is between 50 and 70 dB [28]. Therefore, as illustrated in Fig. 4.2, in both the traditional and modified N-path filters, the IM2 product can be generated due to the downconversion to dc by the mixer, which coincides with the wanted signal. However, the CS-BPF does not experience any frequency translation, thus no IM2 products.

As an application example of such a BPF, the feedback-based superheterodyne RX utilizing a charge-sharing (CS) technique and N-path notch filter was proposed in [6]. Although the N-path notch filter is used as a channel select filter, the N-path folding is of no concern there due to the strong protection offered by the preceding high-IF CS filters. Moreover, in [114], a complete fully integrated superheterodyne RX using the CS technique and a BB filtering was proposed. The folding due to the lower sampling frequency of the BB filters is also of no real concern as it is protected by the preceding high-IF CS filters.

4.3 Charge-Sharing Band-pass Filter (CS-BPF)

The block diagram of the superheterodyne RX front-end is shown in Fig. 4.3. The RF signal of f_{RF} frequency is converted to current, I_{RF}, via a low-noise transconductance

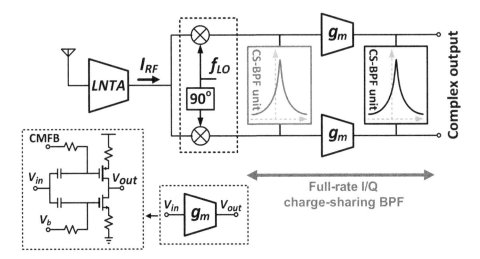

Figure 4.3 Block diagram of the high-IF receiver containing the novel BPF and schematic of IF g_m-cell.

amplifier (LNTA). Then, I_{RF} is downconverted to an intermediate frequency f_{IF} current I_{IF} by a passive mixer comprising commutating switches clocked at f_{LO} rate with rail-to-rail 25% duty cycle. The $f_{IF} = |f_{LO} - f_{RF}|$ frequency could be in the 1–100 MHz range. However, to avoid the unnecessary increase in the power of IF circuitry, f_{IF} should be placed just beyond the flicker noise corner of the devices comprising the RX circuitry [6]. Mixers driven by the 25% duty cycle clocks have a higher conversion gain from RF to IF and also introduce less flicker noise compared to counterparts driven by the 50% duty cycle clock [22]. Hence, this justifies our choice of the double-balanced mixer driven by the 25% clock.

The downconverted I_{IF} current flows into a complex full-rate I/Q CS-BPF. Multiple unit filters of first-order could be cascaded to get high-Q BPF centered at f_{IF}. The discussed filter provides enhanced RX selectivity and rejects unwanted blockers and images inherent to the high-IF architecture.

4.3.1   BPF Unit Structure

The well-known real-valued DT IIR low-pass filter (LPF) is shown in Fig. 4.4(a) [121]. The input charge packet is the integrated input current (provided by a g_m-cell) on C_H and C_R during ϕ_1 over a time window T_s. At ϕ_1 going inactive, C_R samples a portion, $C_R/(C_R + C_H)$, of the integrated input charge. As a result, the DT circuit shown in Fig. 4.4(a) has a first-order DT IIR characteristic, with C_R acting as a lossy component ("switch-capacitor resistor") that leaks the total charge out of the system. Therefore, it prevents the C_H voltage from overflowing, thus ensuring stability. The order of the Fig. 4.4(a) DT IIR filter can be further increased to second or fourth, as shown in Fig. 4.4(b) and Fig. 4.4(c), respectively. At the end of ϕ_1, the sampled charge on C_R

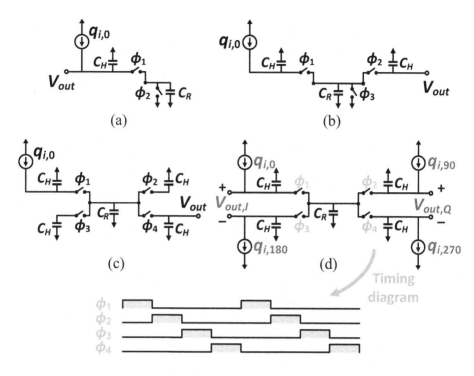

Figure 4.4 Basic concepts of DT charge-sharing IIR filtering: (a) first-order real-valued LPF filter, (b) second-order real-valued LPF filter, (c) fourth-order real-valued LPF filter, and (d) first-order complex-valued BPF filter.

is just shared with another C_H capacitor. This mechanism can arbitrarily increase the IIR filter's order [5].

The basic quadrature (i.e., with four outputs) CS-BPF can be synthesized from the fourth-order DT IIR filter (with a single real output) by applying input charge packets $q_{i,0}$, $q_{i,90}$, $q_{i,180}$, and $q_{i,270}$ with a multiple of 90° phase shifts, as shown in Fig. 4.4(d). During each phase of ϕ_1, ϕ_2, ϕ_3, and ϕ_4, four input charge packets are accumulated into their respective history capacitors, C_H. At the end of each phase, each C_R containing the previous packet is ready to be charge shared with C_H containing the current input charge packet and the "history" charge. Therefore, in each phase, rotating capacitor C_R removes a charge proportional to $C_R/(C_H + C_R)$ from each C_H and then delivers it to the next C_H. The four quadrature outputs can be read out at the sampling rate of $f_s = 1/T_s = f_{LO}$. In that case, The CS-BPF is not full-rate anymore and its sampling frequency would be equal to f_{LO}.

The basic concept of the I/Q charge-sharing filtering with *active* opamps was introduced in [122] for a different low-IF application with very low sampling rate of 1 Msample/s. In our work, the 5.2 Gsample/s CS-BPF is fully passive without any opamps, constructing DT filters that are much more robust to mismatches than the RC, LC, and G_m-C type of filters because of the excellent capacitor matching in advanced CMOS. The other advantage of this filter is that it is fully compatible with process scaling due to the filter's passive nature.

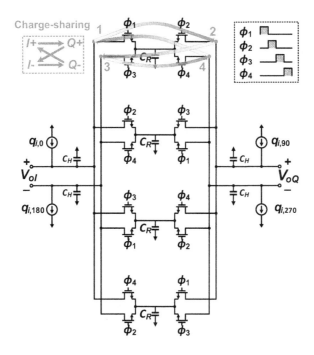

Figure 4.5 Complex CS-BPF unit circuit.

The first-order CS-BPF unit is shown in Fig. 4.5. The time-domain I/Q output voltage expressions at $t = nT_s$ can be written as

$$V_{oI}[n] = \frac{C_H V_{oI}[n-1] - C_R V_{oQ}[n-1] + q_{in,I}[n]}{C_H + C_R}, \tag{4.1}$$

and

$$V_{oQ}[n] = \frac{C_H V_{oQ}[n-1] + C_R V_{oI}[n-1] + q_{in,Q}[n]}{C_H + C_R}. \tag{4.2}$$

By defining the complex input charge as $q_{in,C} = q_{in,I} + j q_{in,Q}$ and complex output voltage as $V_{oC} = V_{oI} + j V_{oQ}$, the z-domain complex transfer function of the filter can be derived as

$$H_{CS-BPF}(z) = \frac{V_{oC}(z)}{q_{in,C}(z)} = \frac{k}{1 - (a + j(1-a))z^{-1}}, \tag{4.3}$$

where $k = 1/(C_H + C_R)$, $a = C_H/(C_H + C_R)$. The position of CS-BPF complex pole is determined by a. According to (4.3), the charge-sharing technique forms a first-order complex filter. The ideal transfer functions of the filter for different a coefficients are shown in Fig. 4.6. The CS-BPF is acting as a LPF centered at dc in the extreme case of $a = 1$, while for the extreme case of $a = 0$, CS-BPF is acting as an N-path

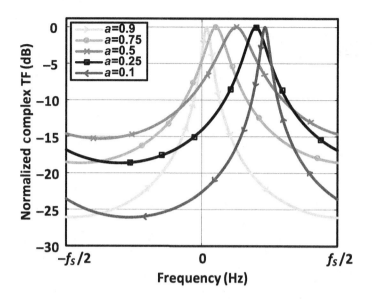

Figure 4.6 Ideal CS-BPF transfer function.

filter centered at $f_s/4$. Moreover, the filter bandwidth increases when $a < 0.5$, and decreases when $a > 0.5$, with the increase of the center frequency f_c.

As C_H increases, the center frequency (f_c) of the transfer function moves toward zero where the BPF becomes more of a low-pass filter (LPF), similar to that in [10]. By increasing C_R, charge sharing between I & Q increases and shifts f_c away from zero toward $f_s/4$. In addition, it reduces the passband gain. In the case where C_H is zero, the BPF turns to a complex N-path filter ($N = 4$) with center frequency at $f_s/4$ (i.e., same as f_{LO}), similar to the one in [116]. The center frequency could be flipped to negative by reversing the sequence of clock signals (ϕ_{4-1}). As also calculated in [122], f_c derived from (4.3) is

$$f_c = \frac{f_s}{2\pi} \arctan\left(\frac{C_R}{C_H}\right). \tag{4.4}$$

This shows that the transfer function of this filter is set only by the sampling frequency and C_R/C_H capacitor ratio. Therefore, it has little sensitivity to process, voltage, and temperature (PVT) variations. Passband gain of this filter is calculated by replacing $z = e^{j2\pi f T_s}$ in (4.3) at the center frequency (4.4):

$$A_{BPF} = \frac{1}{C_R + C_H\left(1 - \sqrt{1 + (C_R/C_H)^2}\right)} \approx \frac{1}{C_R} \ \text{for} \ C_R \ll C_H. \tag{4.5}$$

As in practical applications, f_s is in the GS/s range while IF (and so f_c) is around tens of MHz; the assumption that C_R is much smaller than C_H is reasonable. The unit of (4.5) is volt/coulomb.

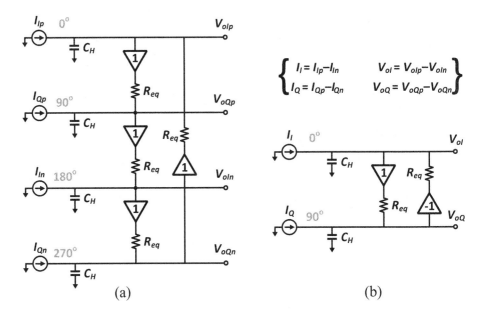

Figure 4.7 Schematics of the continuous-time model of quadrature DT CS-BPF with: (a) single-ended and (b) differential inputs.

4.3.2   CS-BPF Continuous-Time Model

The switched-capacitor circuit of CS-BPF can be modeled as an RC network for frequencies of interest below $f_s/10$. The continuous-time (CT) equivalent model of the DT CS-BPF is shown in Fig. 4.7 for (a) single-ended and (b) differential inputs. Phase of input currents (I_{Ip}, I_{Qp}, I_{In}, and I_{Qn}) should be 0°, 90°, 180°, and 270°, respectively that can be generated with the conventional quadrature current-commutating passive mixer. R_{eq} is an equivalent DT resistance of C_R and is equal to $1/(C_R f_s)$. The input currents are integrated into C_H's, and the charge-sharing with C_R's is modeled with R_{eq} isolated by a unity-gain buffer to account for DT time-division duplexing (TDD) isolation between the quadrature paths. The CT transfer functions (TF) of Fig. 4.7(a) and (b) are ultimately the same. Since the differential input interpretation reduces the number of expressions to half, the differential TF analysis will be carried out below. The s-domain voltage–current expressions of the Fig. 4.7(b) circuit can be written as

$$V_{oI}(s) = I_I(s) \cdot \frac{R_{eq}}{1 + s R_{eq} C_H} - V_{oQ}(s) \cdot \frac{1}{1 + s R_{eq} C_H}, \tag{4.6}$$

and

$$V_{oQ}(s) = I_Q(s) \cdot \frac{R_{eq}}{1 + s R_{eq} C_H} + V_{oI}(s) \cdot \frac{1}{1 + s R_{eq} C_H}. \tag{4.7}$$

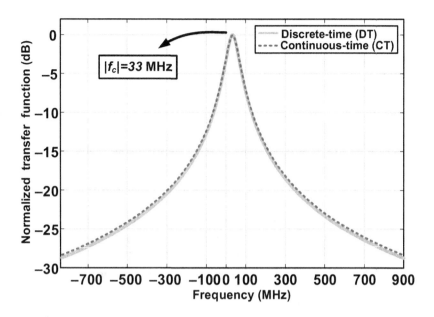

Figure 4.8 Transfer function comparison between the discrete-time CS-BPF and its continuous-time model.

By defining a differential complex output as $V_{oC}(s) = V_{oI}(s) + jV_{oQ}(s)$, and differential complex input current as $I_{in,C}(s) = I_I(s) + jI_Q(s)$, the complex s-domain transfer function of the CS-BPF can be derived from (4.6) and (4.7) as

$$H(s)|_{s=j\omega} = \frac{V_{oC}(s)}{I_{in,C}(s)} = \frac{R_{eq}}{1 - j(1 - R_{eq}C_H\omega)}. \tag{4.8}$$

Consequently, the center frequency of the introduced CT-models in Fig. 4.7 is at

$$f_c = \frac{1}{2\pi R_{eq}C_H}, \tag{4.9}$$

and the complex input impedance is equal to R_{eq}. Moreover, the bandwidth of the CS-BPF can be found from (4.8) and (4.3), which is equal to $1/(\pi R_{eq}C_H)$ for $a \approx 1$. Therefore, there is always a direct relationship of $f_c \approx BW/2$ for $a \approx 1$. It should be mentioned that (4.8) can be derived from (4.3) by performing a bilinear transformation with an approximation of $sT_s < 2$ and substituting $z = (2+sT_s)/(2-sT_s)$ and $s = j\omega$ into (4.3). As an example, for a CS-BPF with $C_R = 1\,\mathrm{pF}$, $C_H = 19\,\mathrm{pF}$ and $f_s = 4\,\mathrm{GHz}$, we find $R_{eq} = 250\,\Omega$ and $f_c = 33.5\,\mathrm{MHz}$. The corresponding DT and CT transfer functions are plotted in Fig. 4.8 and show excellent agreement.

4.4 Noise Analysis of CS-BPF

The total output noise of the CS-BPF contains the noise of all switches within the passive switched-capacitor network. At first, let us analyze the noise of the simplest switched-capacitor circuit in Section 4.4.1. Afterward, the detailed noise analysis of the CS-BPF will be described for DT/CT model in Section 4.4.2.

4.4.1   Voltage Sampler Output Noise

A voltage sampler that includes the noise of its switch is drawn in Fig. 4.9(a). Let us assume that V_{in} is zero. When the switch is turned on, it has a finite resistance R_{on}. A series voltage source models the resistor's thermal noise with a constant power spectral density (PSD), as shown in Fig. 4.9(b).

$$S_R(f) = 4kT R_{on} \qquad f \geq 0, \tag{4.10}$$

where k is Boltzmann constant and T is the absolute temperature. When the switch is on, noise of the resistor is shaped by the RC filter with a time constant of $\tau = R_{on} C_R$ and then appears at the output. At the moment the switch is disconnected, the output noise is sampled and held on C_R. The periodical sampling at f_s causes noise folding from frequencies higher than $f_s/2$, to the 0-to-$f_s/2$ range, where they add up, as shown in Fig. 4.9(c). If the time constant τ is much shorter than the turn-on time of the switch, it can be shown that the summation of all folded noise will be flat

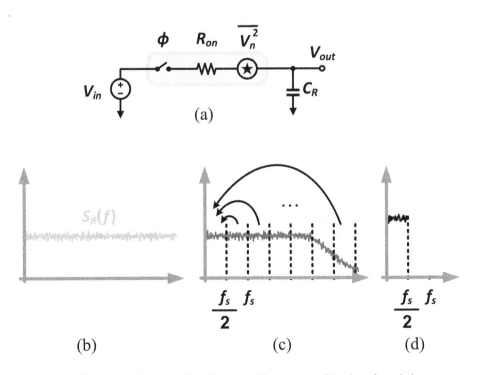

(a)

(b)   (c)   (d)

Figure 4.9 (a) Noise circuit model of a voltage-sampling process, (b) noise of a switch resistance, (c) noise shaped by RC filter, and (d) sampled noise.

(i.e., white noise) [109]. As shown in Fig. 4.9(d), the single-sided noise spectral density of the sampled output noise is [109]

$$\overline{v_n^2}(f) = \frac{kT}{C_R f_s/2}, \qquad 0 \le f \le f_s/2. \tag{4.11}$$

It should be noted that the integrated power density of this noise over the entire frequency range is kT/C_R.

To simplify the calculations for more complicated switched-capacitor circuits, we can make the following assumption: the continuous-time noise source with PSD of (4.10), can be considered as a discrete-time noise source with PDS described in (4.11). In this way, it is not necessary anymore to consider the effect of RC filtering.

4.4.2   DT CS-BPF Noise Model

The simplified noise model of CS-BPF for only one C_R is shown in Fig. 4.10. The input charge packets are assumed zero and the switches are assumed ideal. The first purpose of the following calculations is to find the DT output noise levels $V_{oIp}, V_{oQp}, V_{oIn}$, and V_{oQn} generated by input noise sources $\overline{V_{n1}^2}, \overline{V_{n2}^2}, \overline{V_{n3}^2}$ and $\overline{V_{n4}^2}$. The second purpose is to find the total pseudo-differential output noise of I or Q paths in both DT and CT models. The aforementioned input noise sources have two conditions: (1) they are uncorrelated, and (2) the stochastic value of each of them is

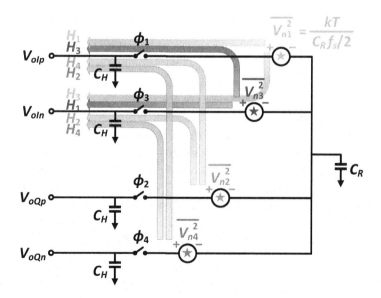

Figure 4.10 CS-BPF noise model for only one of the switches.

equal to (4.11). We first assume that $\overline{V_{n2}^2}$, $\overline{V_{n3}^2}$, and $\overline{V_{n4}^2}$ are zero to calculate the noise transfer function only from $v_n = \sqrt{V_{n1}^2}$ to all outputs. The time-domain noise outputs at $t = nT_s$ with respect to the input noise source $v_n[n]$ can be written as

$$V_{oIp}[n] = a V_{oIp}[n-1] + b V_{oQn}[n-1] + b v_n[n], \tag{4.12}$$

$$V_{oQp}[n] = a V_{oQp}[n-1] + b V_{oIp}[n-1] - b v_n[n-1], \tag{4.13}$$

$$V_{oIn}[n] = a V_{oIn}[n-1] + b V_{oQp}[n-1], \tag{4.14}$$

and

$$V_{oQn}[n] = a V_{oQn}[n-1] + b V_{oIn}[n-1]. \tag{4.15}$$

where $a = C_H/(C_H + C_R)$, and $b = 1 - a$ are the same as before. By converting the time-domain expressions to z-domain, we find DT noise transfer functions as,

$$H_1 = \frac{V_{oIp}}{v_n} = -\frac{b(1 - az^{-1})^3 - b^3 z^{-4}}{b^4 z^{-4} - (1 - az^{-1})^4}, \tag{4.16}$$

$$H_2 = \frac{V_{oQp}}{v_n} = \frac{a(1 - az^{-1})^3(1 - z^{-1})}{b^4 z^{-4} - (1 - az^{-1})^4} \cdot \left(\frac{bz^{-1}}{1 - az^{-1}}\right), \tag{4.17}$$

$$H_3 = \frac{V_{oIn}}{v_n} = \frac{a(1 - az^{-1})^3(1 - z^{-1})}{b^4 z^{-4} - (1 - az^{-1})^4} \cdot \left(\frac{bz^{-1}}{1 - az^{-1}}\right)^2, \tag{4.18}$$

and

$$H_4 = \frac{V_{oQn}}{v_n} = \frac{a(1 - az^{-1})^3(1 - z^{-1})}{b^4 z^{-4} - (1 - az^{-1})^4} \cdot \left(\frac{bz^{-1}}{1 - az^{-1}}\right)^3. \tag{4.19}$$

The above expressions are derived based on the assumption of $\overline{V_{n2}^2}$, $\overline{V_{n3}^2}$, and $\overline{V_{n4}^2}$ being zero. It should be mentioned that, since the circuit is symmetric for all four input noise sources in Fig. 4.10, the noise TF of other DT input noise sources to the output combinations are exactly the same as (4.16)–(4.19). The only difference is that the outputs in the expressions should be changed according to the DT input noise sources; for instance, the noise TF of $\sqrt{V_{n3}^2}$ to V_{oIn} is the same as (4.16). The detailed noise TF for each DT input noise is also illustrated in Fig. 4.10. To calculate a differential DT output noise ($V_{on} = V_{oIp} - V_{oIn}$) with respect to all four input noise sources, we should consider that the differential DT output noise is composed of a sum of four uncorrelated noise contributions, as shown in Fig. 4.10. Moreover, each of them has two correlated noise contributions in the differential output. The correlated noises are shown with the same color (see Fig. 4.10). Therefore, we find the DT differential output noise PSD as

$$\overline{V_{on}^2} = \left|(H_1 - H_3)^2\right| \overline{V_{n1}^2} + \left|(H_4 - H_2)^2\right| \overline{V_{n2}^2} + \left|(H_3 - H_1)^2\right| \overline{V_{n3}^2} + \left|(H_2 - H_4)^2\right| \overline{V_{n4}^2}. \tag{4.20}$$

Since the absolute values of the four input sources are the same, (4.20) can be simplified as

$$\overline{V_{on}^2} = \left(2\left|(H_1 - H_3)^2\right| + 2\left|(H_2 - H_4)^2\right|\right) \cdot \overline{V_{n1}^2}, \tag{4.21}$$

$$\overline{V_{on}^2} = \frac{2\,b^2\left(\left(\cos\left(\frac{w}{f_s}\right)\right)^2 b - a\cos\left(\frac{w}{f_s}\right) + a^2\right)}{\left(b^2 + a^2\right)\left(\cos\left(\frac{w}{f_s}\right)\right)^2 + \left(2\,b^3 - 4\,b^2 + 4\,b - 2\right)\cos\left(\frac{w}{f_s}\right) + a^2\left(b^2 + 1\right)}$$

$$\cdot \left(\frac{kT}{C_R f_s/2}\right). \tag{4.22}$$

and by substituting $z = e^{j\omega/f_s}$, the differential output noise PSD is simplified to (4.22). The comparison of calculated output noise PSD based on (4.16)–(4.19) with transistor-level simulations are illustrated in Fig. 4.11, for $C_R = 4\,\text{pF}$, $C_H = 19\,\text{pF}$, and $f_s = f_{LO} = 1\,\text{GHz}$. The differential output noise PSD of the CT model of Fig. 4.7 can be calculated based on the same approach; DT noise PSD derived in (4.22). We find the total CT differential output ($V_{oIp} - V_{oIn}$) noise PSD as

$$\overline{V_{on}^2}(\omega) = \left(\frac{2(R_{eq}C_H\omega)^2 + 4}{(R_{eq}C_H\omega)^4 + 4}\right) \cdot (4kT R_{eq}). \tag{4.21}$$

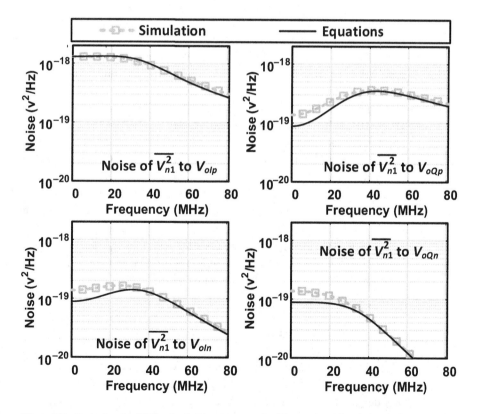

Figure 4.11 Output noise PSD calculations compared with transistor-level simulations.

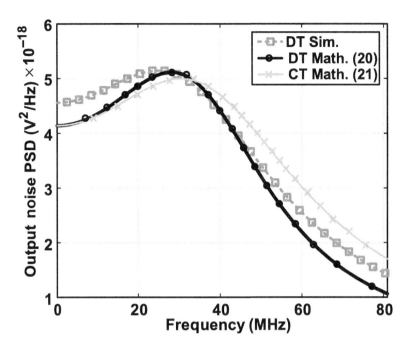

Figure 4.12 Output noise PSD calculations compared with simulations.

It should be pointed out that integrating the DT differential output noise PSD in (4.22) over 0-to-$f_s/2$ yields kT/C_T, with C_T being the total differential output capacitance equal to $(C_H + C_R)/2$. On the other hand, integrating the CT noise PSD in (4.21) over the entire range of 0 to ∞ is again equal to kT/C_T, with $C_T = C_H/2$. Note that the unity gain buffers in Fig. 4.7 are merely conceptual to account for the DT isolation, hence noiseless. If one were to implement the CT circuit of Fig. 4.7, noise contribution of the buffers would have to be accounted for. Consequently, the DT CS-BPF of Fig. 4.5 has a potential to outperform its CT counterpart.

As the final verification, Fig. 4.12 compares the total output spot noise plots obtained via the diverse means: calculated DT, based on (4.22); calculated CT, based on (4.21); and schematic-simulated DT. The following conditions are used: $C_R = 4\,\text{pF}$, $C_H = 19\,\text{pF}$, and $f_s = f_{LO} = 1\,\text{GHz}$. Although all simulations and calculations are performed for the CS-BPF with one C_R, the presented approach is valid for the full-rate CS-BPF with only one difference: the f_s in full-rate CS-BPF is four times higher than CS-BPF with one C_R.

4.5   Circuit Implementation

To accurately measure the BPF linearity, we have replaced the LNTA with a simple self-biased inverter-based transconductance amplifier ($g_{m,RF}$) for higher IIP3, and designed for small transconductance as not to degrade the linearity. Since the gain provided by $g_{m,RF}$ is small, its contribution to the input-referred noise (IRN) is

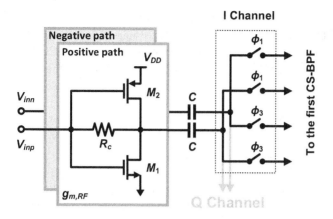

Figure 4.13 Circuit implementation of $g_{m,RF}$ and mixer.

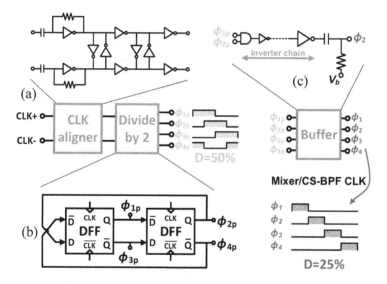

Figure 4.14 Schematic of the clock generation circuit: (a) CLK aligner circuit, (b) divider, and (c) 25% clock generation circuit with buffer stage.

predominant. The schematics of $g_{m,RF}$ and RF mixer are shown in Fig. 4.13. The self-biasing of $g_{m,RF}$ is accomplished by R_c resistors connecting its input and output. The value of R_c in parallel with the output impedance of $g_{m,RF}$ should be high enough as not to degrade the Q of first CS-BPF. The dc block capacitors (C) are used to eliminate the dc current flowing into CS-BPF. The differential RF input voltage to $g_{m,RF}$ is converted to a pseudo-differential AC current feeding the commutating CMOS passive mixers of I and Q channels. The RF mixer in Fig. 4.13 is only shown for the I channel.

The clock phases ϕ_1 and ϕ_3 comprise a pseudo-differential 25% duty cycle (D) LO clock driving the CMOS switches. Figure 4.14 presents the clock generation circuit for both the mixer and CS-BPF. The differential input clock, CLK, with $D = 50\%$ is applied to the aligner circuitry that is responsible to compensate for any phase mismatch between the CLK+ and CLK− differential phases. The CLK aligner circuit (see Fig. 4.14(a)) consists of two inverters at the input to convert the sinusoidal inputs to the square-wave clock with $D = 50\%$ and the two stages of back-to-back inverters for further aligning the complementary edges of the square-wave clock.

As shown in Fig. 4.14(b), the divide-by-2 circuit consists of two D flip-flops arranged in the loop to generate the $D = 50\%$ clocks, ϕ_{1p}, ϕ_{2p}, ϕ_{3p}, and ϕ_{4p}, with 25% delay between adjacent edges. The mixer clock is generated by the buffer shown in Fig. 4.14(c). The CS-BPF switches are driven by the clocks generated in another buffer with the same schematic as drawn in Fig. 4.14(c). It comprises AND gates and the chain of inverters for proper driving of the load capacitance of NMOS switches. Moreover, to increase the driving capability of sampling switch transistors in the quadrature mixer and CS-BPF, a clock boosting technique (using V_b, see Fig. 4.14(c)) is utilized to increase gate-source voltage while the pass transistor is turned on.

The CS-BPF operates at clock frequency f_{LO} with 25% duty cycle clocks and its effective (i.e., differential I/Q) sampling frequency f_S is equal to $4f_{LO}$. Thus, the effective sampling time T_S is equal to $1/(4f_{LO})$. To maximize the linearity, it is crucial to set the switch sizes of Fig. 4.5 in such a way that T_S would be between 3τ and 4τ. τ is the $R_{on}C_R$ time constant of the DT circuit, and R_{on} is an equivalent resistance of the sampling transistor in the triode region. The output resistance of the IF g_m-cell should be at least $3\times$ higher than R_{eq} in order to not decrease the Q and bandwidth of the following CS-BPF.

4.6   Measurement Results

The introduced RX with the same structure as Fig. 4.3 but with three-stage CS-BPF together with its surrounding circuitry was fabricated in TSMC 1-poly and 7-metal layers 65 nm CMOS. The chip micrograph is shown in Fig. 4.16. The implemented RX occupies 0.45 mm^2 active area and consumes 24.5 mA at 1.2 V power supply.

4.6.1   Test Setup

The introduced front-end IC is wire-bonded to a printed circuit board (PCB) providing dc and RF input connectivity ports, while high-IF output signals are measured with a high performance oscilloscope, as shown in Fig. 4.17. The transfer function measurement setup of the RF CS-BPF is shown in Fig. 4.17(a). After providing the proper power supply voltages, the LO frequency and RF input frequency should be applied to the RF front-end. The quadrature (I/Q) IF outputs are connected to a high-performance "RTO 1044" digital oscilloscope, and the process of taking fast fourier transform (FFT) from the IF output signals has performed with its own digital oscilloscope

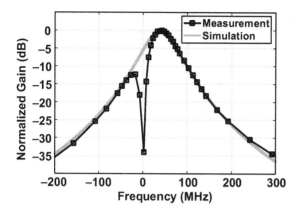

Figure 4.15 Comparison of measured transfer function with an ideal transfer function that includes output impedance of g_m-cells.

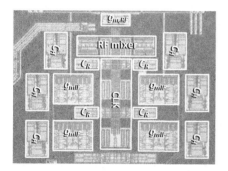

Figure 4.16 Chip micrograph.

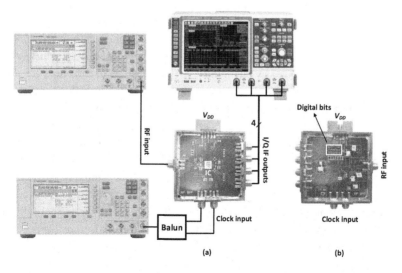

Figure 4.17 (a) Measurement setup and PCB of the introduced front-end for top layer, (b) PCB bottom layer.

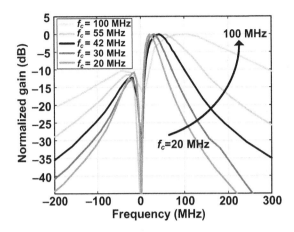

Figure 4.18 Measured transfer function for different IF frequencies. Center frequency f_c aligns with f_{IF}.

software GUI. The IF g_m-cell gains and the values of C_H and C_R capacitors controlled by the DIP switch are shown in Fig. 4.17(b).

The measured complex transfer function of the RX is shown in Fig. 4.15. The measured curve is also compared to an ideal mathematic transfer function that includes the output impedance of all g_m-cells, which was extracted from transistor-level simulations. The measured curve shows a very good agreement with the mathematic modeling except for a notch at dc. It is due to the high-pass characteristic of a dc block capacitor in the g_m-cell (see Fig. 4.3) together with the resistor, providing bias and common-mode voltages.

To demonstrate the CS-BPF reconfigurability, the measured transfer functions for different center frequencies f_c and bandwidths are depicted in Fig. 4.18. The stopband rejection of the filter improves with f_c and without any replicas as in the conventional architectures (see Fig. 4.1). The measured center frequencies of the transfer functions are controlled by changing C_H (see (4.9)). C_H capacitors are implemented as a digitally switchable binary weighted capacitor using the conventional MOM capacitors and MOS switches. Hence, the C_H value can be changed via 6 digital bits.

The complete front-end provides a total gain of 35 dB at the maximum gain setting. The measured and simulated IRN of the front-end are shown in Fig. 4.19. The abrupt increase in IRN at low frequencies is caused by the flicker noise of the g_m-cell at IF stage. As discussed in Section 4.3, this curve suggests that the IF frequency should be placed at 30 MHz or a bit higher. In addition, the reason that the measured IRN is high is that the front-end ($g_{m,RF}$ and first CS-BPF) gain is low not to sacrifice the linearity of the RX. As a consequence, the higher IRN is measured.

As shown in Fig. 4.20, the out-of-band IIP3 of the RF front-end ("$g_{m,RF}$+first CS-BPF") is measured by applying a two-tone at the input of the chip. The out-of-band two-tone frequencies are at 1100.009 MHz, 1200 MHz to have enough filtering at the output of RF front-end for reducing the linearity contribution of the rest of

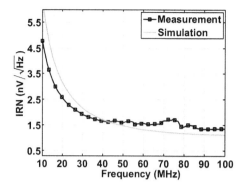

Figure 4.19 Measured and simulated IRN for $C_H = 10\,\mathrm{pF}$ and $C_R = 1\,\mathrm{pF}$.

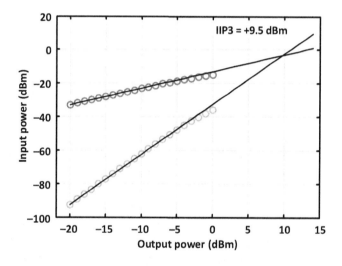

Figure 4.20 Measured out-of-band IIP3 of the RF front-end ($g_{m,RF}$ + first CS–BPF).

the RX chain. The measured IIP3 is $+9.5\,\mathrm{dBm}$, and we believe that the measured IIP3 is chiefly limited by the linearity of the $g_{m,RF}$-cell because the simulated IIP3 of the CS-BPF itself is more than $+30\,\mathrm{dBm}$. Table 4.1 shows summary of the filter and compares it with state-of-the-art. Compared with other designs except [117], the power consumption of our test chip is lower, but the filter order of our test chip is two orders higher than [117]. Compared to [118], the power consumption of our test chip is almost half for the highest sampling frequency. Moreover, CS-BPF provides higher reconfigurability, and wider BW selectivity of 24–125 MHz. In addition, it has a digitally controllable IF center frequency range of 20–100 MHz larger than $1/f$ corner frequency, unlike other filters [117, 118]. Although the input g_m-cell has degraded the linearity of the test chip, the in-band and out-of-band IIP3 of 0 dbm and $+10\,\mathrm{dBm}$ is achieved, respectively.

Table 4.1. Summary and comparison with state-of-the-art

	This work	[8]	[117]	[118]
CMOS Tech. (nm)	65	65	65	65
Type	Filter	Receiver	Filter	Filter
V_{DD} (V)	1.2	1.2/2.5	1.2	1.2
Power (mW)	28	39	2–20	18–57
IRN (nV/$\sqrt{Hz}$)	1.5	0.87	0.9-1.3	0.87
IB-IIP3 (dBm)	0	N.A	N.A	−12
OB-IIP3 (dBm)	+9.5	N.A	+14	+26
BW (MH)	24–125	4	35	8
Filter order	6	6	2	6
IF Freq. (MHz)	20–100	62	–	–
Freq. range (GHz)	0.5–1.2	1.8–2.2	0.1–1	0.1–1.2
Active area (mm^2)	0.19	0.76	0.07	0.27

4.7   Conclusion

Process-scalable fully integrated band-pass filters (BPF), free from replicas to be suitable for high-IF or superheterodyne receivers (RX) are in high demand to solve the issues related to continuous-time (CT) and N-path filters. We introduce and analyze a discrete-time (DT) charge-sharing (CS) BPF that is entirely passive and uses transistors only as switches. The center frequency of the introduced BPF filter is digitally controllable via clock frequency and capacitor ratio and thus insensitive to PVT variations. It is free from aliasing and replicas while operating at a GSample/s rate. The filter performance is verified in 65 nm CMOS for the wide RF frequency range of 0.5–1.2 GHz and a digitally controllable center frequency of 20–100 MHz. Measured noise performance and transfer function of the filter accurately fit both the mathematical theory and the CT schematic model. The experimental results indicate the introduced filter to be a prime candidate for superheterodyne receivers.

5 Discrete-Time Receivers: Case Studies

In this chapter, we describe four realized examples of discrete-time receivers that are largely based on the architecture and circuitry introduced in Chapters 3 and 4. We start with a commercial DT receiver designed for GSM single-chip radios, which introduces the novel low-pass IIR filter, then continue with three highly reconfigurable superheterodyne receivers that employ the complex IQ charge-sharing band-pass filter (BPF) for image rejection.

5.1 GSM Receiver Front-End Architecture in 90 nm CMOS

The first-ever DT receiver front-end for wireless cellular radios is shown in Fig. 5.1. It was mass-produced by Texas Instruments. The RX consists of an LNA followed by two transconductance amplifiers (TA) and two passive mixers. The RF input signal is amplified by the LNA and splits into I/Q paths where it is further amplified by the TA. It is then downconverted to a low intermediate frequency (IF) that is fully programmable (but defaults to 100 kHz) by the following mixers driven by an integrated local oscillator (LO). The IF signal is sampled and low-pass filtered by passing through the switched-capacitor-filter (SCF). The LO signals are generated using an all-digital PLL (ADPLL) [123] that incorporates a digitally controlled oscillator (DCO). The digital control unit (DCU) provides all the clocks for the SCF operation.

Although the front-end circuit requires two TAs, two mixers, and quadrature LO signals, the receiver has an excellent sensitivity and good linearity at a low supply voltage (V_{DD}) of 1.4 V, thus offering excellent performance that satisfies the GSM requirements. The power is supplied by an integrated low-dropout (LDO) regulator.

5.1.1 Low-Noise Amplifier

A differential LNA is implemented to improve the noise figure from substrate coupling originated from DBB since the impact of the switching noise of more than a million digital gates on the same silicon die could not have been known precisely. Figure 5.2 shows a simplified schematic diagram of the LNA. A variable gain feature with seven digitally configurable steps is implemented. In the high gain mode, four voltage gains are realized with a 2 dB step between 21 dB and 29 dB. In the low-gain mode, there

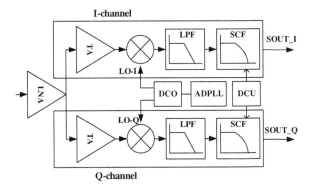

Figure 5.1 GSM receiver front-end diagram.

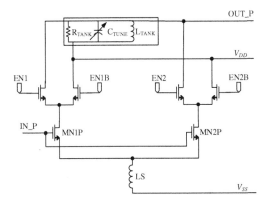

Figure 5.2 LNA core schematic.

are three gain steps with a 2 dB step between 3 dB and 9 dB. As shown in Fig. 5.2, the multiple cascode stages are connected in parallel with one source degeneration inductor and one inductive load. Each stage has digital configurability.

The top transistors of the cascode stage used for bypassing the gain contribution are shunted to V_{DD}. Since the bottom transistors of the cascode stage operate in all gain settings, the input impedance is constant to the first order over the gain selection, which is critical for constant input power and noise matching. Inductive source degeneration using package bond wires is implemented to improve linearity. The LNA load is an on-chip spiral inductor using multiple metal layers with metal width = 5.9 µm, metal space = 2 µm, inner diameter = 81.9 µm and 10 turns. This inductor is drawn as a center-tap configuration for better matching between the differential branches and for achieving a higher quality factor (Q). As shown in Fig. 5.3, the inductance is 8.9 nH and Q is >4 at 900 MHz, where Q is defined as $|imag(y11)/real(y11)|$. To reduce the substrate effect, all doping under the inductor is blocked to preserve a higher resistivity.

The inductor is tuned with the capacitance at the LNA load, which comprises tuning capacitors together with parasitics. The tuning capacitor is realized using

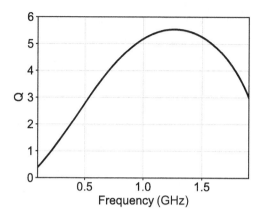

Figure 5.3 Inductor Q-factor.

metal-insulator-metal (MIM) capacitors and switches. Two capacitors are connected differentially with a switch and two pull-down transistors to keep both source or drain voltages of the switch low and Q of the capacitor bank high. The achieved effective Q is 100 at 900 MHz. When the switch is turned off to be in a low capacitance value, the parasitic capacitance of the MIM capacitors and transistors still has an effective Q of $\sim$100. Compared to a MOS capacitor, the MIM capacitor provides a much better trade-off between Q and C_{ON}/C_{OFF} ratio. In this design, a C_{ON}/C_{OFF} ratio of larger than 4 was achieved while Q is still greater than 100. The selectable capacitance ranges 2.5 pF in total because in this process, MIM capacitance can vary up to $\pm 20\%$ from its nominal value. With this design, all GSM bands can be fully covered.

The differential LNA draws 7.3 mA. The LNA input is protected against ESD by one reverse-biased diode to V_{DD} and three forward-biased diodes in series to V_{SS}. The ESD structures at LNA input are aimed to protect larger than 2 kV human body model (HBM). The LNA bond pad is shielded with a lower metal-1 layer to eliminate the substrate coupling while minimizing parasitic capacitance, which is about 100 fF.

5.1.2 TA and Mixer

Figure 5.4 shows a simplified TA and mixer schematic diagram. A highly efficient push-pull amplifier is chosen for the TA because of its low noise and good linearity characteristics. The variable gain feature is implemented in the TA with a 3-bit control. A feedback amplifier is used to set the dc bias voltage of the TA output node to V_{REF}, which is set to half of V_{DD} so as to provide a maximum signal swing. Resistors in Fig. 5.4 are large enough to prevent significant RF signal loading. The differential TA draws 4 mA in the maximum gain mode.

A double-balanced switching mixer is connected to the TA output via AC-coupling capacitors so that the dc voltage at the TA output is isolated from the mixer. This topology has an excellent feature of the reduced $1/f$ noise because there is no dc current

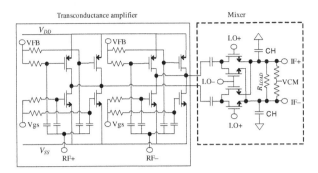

Figure 5.4 TA and mixer core schematic.

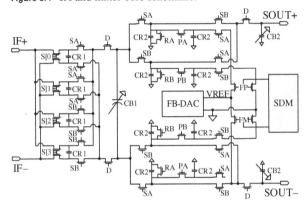

Figure 5.5 SCF core schematic.

flowing through, making it suitable for direct-conversion or near-zero IF receivers. By adding a capacitive load (C_H, history capacitor) to the mixer output, low-pass filtering can be obtained to reduce large interferers. In this mixer, two switches are toggled by one of the complementary LO signals (LO+, LO−) from a digitally controlled oscillator (DCO). Since the mixer is connected to the switched-capacitor filter (SCF), its loading effect can be represented as R_{load} which is about 4.5 kΩ.

5.1.3   Switched-Capacitor Filter (SCF)

The schematic diagram of the switched-capacitor filter block (SCF) is shown in Fig. 5.5. The switches are controlled by the digital control unit (DCU) that generates the timing waveforms shown in Fig. 5.6. For one LO cycle, the RF signal of the mixer output is integrated into a history capacitor (C_H) and a rotating capacitor (C_{R1}). Since the four rotating capacitors sequentially connect to C_H in a fixed order, the charge transfer via C_{R1} is a direct sampling of the IF signal. It is also clear that a charge loss on C_H through C_{R1} creates the loading (R_{load}) on the mixer output. For two LO cycles, two rotating capacitors in the first bank sample the IF signal on C_H while the rotating capacitors in the second bank and C_{B1} share charge. Because of the half-sampling rate from the mixer output to C_{B1}, the decimation operation creates

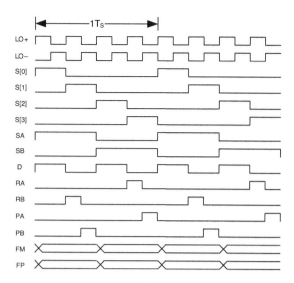

Figure 5.6 DCU clock diagram.

a sinc function that has notches at the foldover frequencies, $NLO/2$, where N is a positive integer. Transconductance (g_m) of TA, C_H, and the loading (R_{load}) of SCF create the first IIR filtering response of g_m-C antialiasing low-pass filtering prior to the main sinc filter. However, the TA sees a periodic constant load at its output.

After the two C_{R1} capacitors in one bank are disconnected from C_H, these carry the charge of past 2 IF samples created by the charge sharing between two C_{R1} and C_H. Next, the two C_{R1} capacitors share charge with the buffer capacitor C_{B1} and a second rotating capacitor, C_{R2}. The overall effect is to create a second IIR filtering stage in which $2C_{R1}$ delivers input; C_{B1} holds the memory; and C_{R2} captures a glimpse of the output of the second IIR filter stage. This charge is subsequently shared with a second buffer capacitor, C_{B2}, resulting in the third IIR filter stage. While charge samples are passed on from C_H to C_{B2} through a series of charge combination, splitting, and recombination operations, the IF information at mixer output are always kept on C_H together with two C_{R1} capacitors from one bank. The three IIR filters have corner frequencies that are given by the respective ratios of rotating capacitors (C_{R1}, C_{R2}) to fixed capacitors (C_H, C_{B1}, C_{B2}) and may be readjusted by changing the size of the capacitors. The capacitor ratios in the SCF are programmable, which allows the filter corner frequency to be adjustable over a wide range, thereby allowing its use in a multistandard environment.

After the charge sharing of C_{R2} with C_{B2}, C_{R2} is reset (RA, RB) and precharged (PA, PB) by the 1-bit feedback circuit (FB-DAC) provided by a sigma-delta modulator that connects the output of a low-noise feedback voltage reference to C_{R2}. Zero DAC code produces approximately 50% duty cycle at FM and FP clocks, which brings the common-mode voltage of the SCF exactly at half of V_{REF}. In the presence of a dc offset, the duty cycle is changed with sigma-delta noise shaping to cancel the offset voltage.

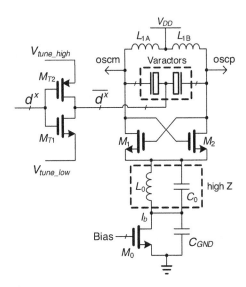

Figure 5.7 DCO core schematic.

5.1.4   Digitally Controlled Oscillator

A DCO circuit schematic is shown in Fig. 5.7 [123]. L1A and L1B are two halves of a center-tap inductor. Because of the short-coming of this 90 nm digital CMOS Cu process that has thin metal interconnects, it is difficult to design an inductor with even a moderate Q. To enhance the Q-factor of the inductor, an Al layer is patterned and connected in parallel with the Cu windings. M_{3-5} plus the Al layer were used to form L_1 while only M_{3-5} layers were used for L_0. The total Cu and Al thickness are only 0.75 μm and 1.0 μm, respectively. The simulated single-ended Q using a $imag(y11)/real(y11)$ definition is 3.6 and 6.7 at 0.9 and 3.6 GHz, respectively. The differential phase stability Q is 3.6 and 10.2 at 0.9 and 3.6 GHz, respectively [124].

The varactor is implemented using an npoly-nwell MOSCAP structure. Extrapolating from the measurement data, the C_{max}/C_{min} ratio is >3 within the ranges of desired gate length L_g and gate width W_g per finger. The resulting total tolerable fixed parasitic capacitance is 720 fF. MOSCAP was chosen because the gate oxide thickness (t_{ox}) is one of the best controlled parameters in this CMOS process, whose corner variation is within ±2.5%. The four different phases of LO driving the I- and Q-mixers in Fig. 5.4 are generated from the DCO frequency that oscillates at $4\omega_0$, where ω_0 is in the GSM band frequencies. A fully digital circuit (i.e., ADPLL) is built around the DCO to adjust its phase and frequency deviations in a negative feedback manner.

5.1.5   Silicon Realization

The presented techniques have been realized in high-volume products by Texas Instruments. Figure 5.8 shows two chip micrographs representing the first and second generation of digital RF processor (DRP^TM), respectively: (1) commercial

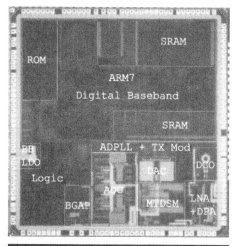

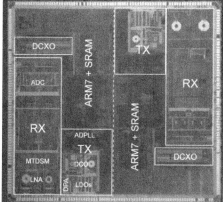

Figure 5.8 Die micrographs of radios employing two generations of DRP: (top) the commercial single-chip Bluetooth radio; (bottom) the pre-production version of the single-chip GSM radio.

10 mm^2 single-chip Bluetooth radio in 130 nm CMOS, and (2) a fully functional pre production version of the single-chip GSM radio in 90 nm CMOS. The GSM chip consists of two independent pairs of transmitters and receivers to study various on-die coupling mechanisms, which are especially important in full-duplex WCDMA operations with RX and TX diversity. The 90 nm process features the following parameters that characterize the process: 0.27 µm minimum metal pitch, five levels of copper metal, 1.2 V nominal transistor voltage, 2.6 nm gate oxide thickness, logic gate density of 250 kgates/mm^2, SRAM cell density of 1.0 Mb/mm^2. The measured RX sensitivity of −82 dBm for Bluetooth and −110 dBm for GSM, versus the respective specifications of −70 dBm and −102 dBm, is quite competitive with conventional solutions. The overall GSM RX noise figure is only 2 dB.

5.1.6 GSM RX Front-End Performance

The LNA input is matched using an external inductor and capacitor with a balun for an impedance ratio of 50 to 100 Ω. The measured LNA input matching with S_{11} is

Table 5.1. Measured performance

	Measured data
Noise figure (1 kHz–100 kHz)	2.0 dB
NF with −25 dBm blocker at 3 MHz offset	5.0 dB
S_{11}	< -10 dB
IIP2	+50 dBm
IIP3	−15 dBm
$P_{1\text{-dB}}$	−25 dBm
GAIN	+34 dB
Front-end current consumption	15.3 mA

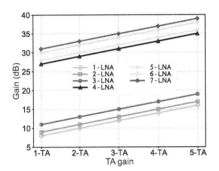

Figure 5.9 Measured voltage gain.

less than −10 dB over the whole GSM band. When the curves of S_{11} versus multiple LNA gains are compared, the largest variation is less than 1 dB. Figure 5.9 displays the front-end voltage gains versus different LNA and TA gain settings. The front-end gains can be configured with an automatic-gain-control (AGC) function to select an optimal gain setting trading off the noise figure for linearity. This circuit adds 32.5 dB dynamic range to the receiver. The measured noise figure in the maximum gain mode is 1.8 dB, which is excellent when considering the fact that several hundred thousand digital logic gates are switching on the same die. With the LO frequency set to 869.1 MHz, +50 dBm of IIP2 is measured with a front-end gain of 34 dB, where the LNA gain is set to 2 dB below the maximum gain (6 LNA) and TA to its middle gain setting (3 TA). To mimic the EDGE environment, two tones of 875.2 MHz and 875.3 MHz are injected into the LNA for the IIP2 test (power of −36 dBm). The LNA, two TAs, and mixers consume 15.3 mA from an internal LDO voltage of 1.4 V. Since this work has a digitally configurable gain with a fine resolution, it is different from a conventional front-end approach that is typically built for two large steps. A major advantage of our approach is that the circuit performance can be finely optimized by selecting the appropriate gain settings.

Table 5.1 summarizes the measured performance when the front-end gain of 34 dB is selected with LNA gain setting number 6 (max −2 dB) and TA gain setting number 3.

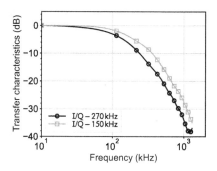

Figure 5.10 Measured filtering characteristics.

Since the SCF is a highly linear filter, little degradation in linearity has been measured. In Fig. 5.10, two pairs of measured plots at SCF output show the low-pass filtering, where the 3 dB frequencies are set to 150 kHz and 270 kHz.

5.2   Feedback-Based Superheterodyne Receiver in 65 nm CMOS

As mentioned in Section 1.2, integrated RF receivers are typically zero-IF or low-IF (i.e., homodyne) because of the well-known benefits, such as high level of integration, the use of low-pass filtering for channel selection, and avoidance of an *external* IF band-pass filter (BPF). Weak desired signals are likely accompanied by large blocking interferers. These blockers can dramatically degrade the receiver performance by causing gain compression and higher-order nonlinearities as well as increasing its noise figure (NF). Conventionally, these out-of-band blockers are filtered out by a bulky and expensive SAW filter placed prior to the LNA input. Since the RF wanted signal could be weak and the dynamic range requirements of a given specification need to be met, the gain of the LNA should be kept high and the blockers should be filtered out. Otherwise, the mixer and the following stages could get saturated. SAW-less receivers have been recently discussed in [125, 126]. They are all based on a homodyne architecture. Unfortunately, they all exhibit well-known homodyne RX issues, such as sensitivity to $1/f$ noise and varying dc offsets, finite IIP2, which will keep on getting worse with the inevitable scaling of the process technology.

In this section, we present a superheterodyne receiver of high-IF that solves the aforementioned issues of the homodyne receivers. Another integrated superheterodyne RX was introduced in [8]. It filters the blockers through an N-path filter, as opposed to the DT filtering approach here. However, the image folding issue was not addressed there. The image folding issues of prior attempts are solved here through a discrete-time (DT) charge-sharing filtering. On the other hand, the blockers are filtered through a feedback-based high-Q RF BPF. The new architecture is process scalable and highly reconfigurable.

The N-path filters offer high-Q BFP filtering with precise control of the center frequency through clock adjustment [8]. Despite a very high-Q filtering, N-path filters provide only around 7–16 dB of filtering rejection due to the poor switch on-resistance in mixers. On the other hand, this type of filter suffers from folding of images from $(N - 1)f_{IF}$ and $(N + 1)f_{IF}$ with a normalized gains proportional to $1/(N + 1)$ and $1/(N - 1)$ [8]. For example, the images of the 16-path filter fold onto the wanted signal via the 24 dB attenuation, which does not appear sufficient. Therefore, it is essential to use prefiltering (i.e., preselect, SAW, N-path) to get rid of the images, which degrade NF and causes image folding. Usually, the gain of LNAs is around 10–20 dB, which can saturate the output of an LNA at a presence of a blocker that can be as large as 0 dBm (600 mV$_{p-p}$). Therefore, to prevent saturation, it is needed to use the BPF right after LNA to attenuate the blockers.

5.2.1   High-Q RF BPF Structure

The novel idea of the high-Q RF BPF comes from a combination of two types of impedances. As shown in Fig. 5.11, the input current is converted to voltage at node X through multiplication by Z_{L1}. Then, it is converted to current and sinked on Z_{L2}. The resulting V_Y voltage gets fed back to input node V_X by a transconductance in the feedback path. As shown in Fig. 5.11, the input impedance of the circuit is $Z_{L1}/(1 + g_{mf}g_m Z_{L1}Z_{L2})$. When the gain $g_{mf}g_m Z_{L1}Z_{L2}$ is smaller than unity, the input impedance is equal to Z_{L1}, which in this design happens at frequencies far from the wanted signal. On the other hand, the input impedance becomes $1/(g_{mf}g_m Z_{L2})$ in the case that $g_{mf}g_m Z_{L1}Z_{L2}$ is much larger than unity, which happens at frequencies very close to the wanted signal. The first impedance Z_{L1} is a third-order complex IQ charge-sharing filter, which acts here as a wide-bandwidth BPF centered at $+f_{IF}$ to filter out images of the wanted signal. The basic concept of IQ charge-sharing filter was introduced in [122] for low-IF with low sampling rate.

The second impedance Z_{L2} is a complex eight-path notch filter (recently introduced in [127] for a real-valued version) to achieve a very sharp high-Q BPF at RF through the feedback path. Figure 5.12 depicts a detailed construction of a low-impedance node after the RF mixer for blockers with an extra filtering at image frequencies. Input matching of the circuit is provided by the input 50 Ω resistance. First, the RF input signal is converted to a current using a simple inverter-based g_m stage followed by a 25 % passive mixer clocked at f_{LO}. The complex output current of the mixer needs a complex low-impedance node for blockers to eliminate the saturation of the

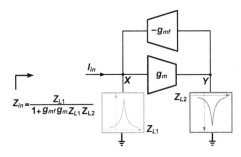

Figure 5.11 Basic concept of impedance combinations.

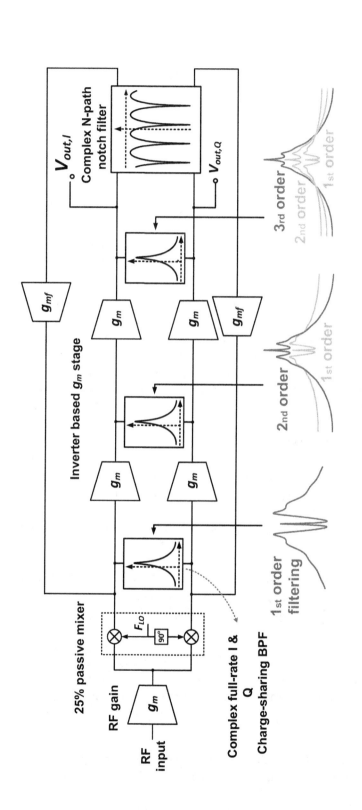

Figure 5.12 Detailed block diagram and operation of the high-IF receiver with high-Q BPF.

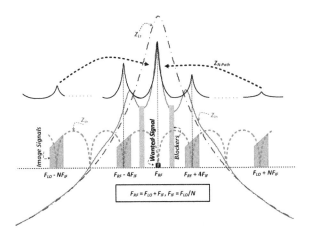

Figure 5.13 Frequency translation of the high-IF receiver compared to a typical N-path filter.

g_m output. As shown in Fig. 5.13, the blockers are attenuated because of the complex high-Q BPF, while the complex full-rate wideband IQ charge-sharing BPF (including the feedback) rejects other image components, including those at $-f_{IF}$. The filtered complex signals go through two similar wideband IQ filters for more attenuation of the images and amplification of the wanted signal. Output signals of the last (third) IQ filter go through a complex notch filter centered at $+f_{IF}$ followed by TIA in order to feed back the complex signals to the mixer output. The notch filter rejects the wanted signal and passes all blockers and unwanted signals, which get fed back through a transconductance (g_{mf}) and will be canceled at the mixer output.

Figure 5.14 shows the concept of IQ charge-sharing wideband BPF. The input current is integrated into the total capacitor $C_t = C_H + C_r$ during the four phases of the nonoverlapping 25% full-rate LO clock. The full-rate operation means that it works at the maximum sampling frequency of $4 f_{LO}$ to avoid decimation. The main drawback of an early decimation would be an unwanted folding due to the change of the sampling rate between stages. Therefore, to avoid aliasing, it is crucial to keep the sampling frequency at the full rate. After each integration of the current into C_t of each quadrature path, a small portion of the total charge $\frac{C_r}{C_t} q_{in}$ is shared between the real and imaginary paths in the next clock cycle. This operation forms a complex filter with a transfer function, which is given by

$$H(z) = \frac{V_{out}(z)}{Q_{in}(z)} = \frac{k}{1 - (a + jb)z^{-1}}, \tag{5.1}$$

where $k = 1/(C_H + C_r)$, $a = C_H/(C_H + C_r)$ and $b = C_r/(C_H + C_r)$. According to (5.1), the charge-sharing process forms a first-order complex filter centered at

$$f_c = \frac{f_s}{2\pi} \arctan \frac{b}{a}. \tag{5.2}$$

Therefore, it is possible to adjust the center frequency f_c by changing the coefficients a and b. However, it is not possible to make the filter very sharp because the DT charge

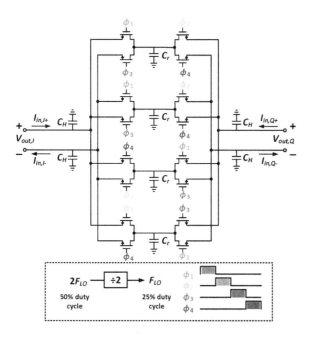

Figure 5.14 Circuit level schematic of the wideband IQ charge-sharing BPF.

sharing is a lossy operation, which increases bandwidth of the filter. f_c is a bit sensitive to the capacitance ratio mismatch, as compared to the N-path filter in which the center frequency is exactly equal to the operating clock frequency. The main advantage of this structure is that the IQ charge-sharing BPF has a very robust filtering at frequencies located at $f_s/2$. As a result, it is feasible to use it as the wideband BPF centered at f_{IF} to reject image signals located at harmonics of f_{IF}. The other benefit of this filter is that its sampling frequency is equal to $f_s = 4f_{LO}$. Therefore, no unwanted folding occurs as compared to the N-path filter, which suffers from harmonics folding.

The presented architecture offers several advantages over the state-of-the-art receivers. The high-IF RX eliminates the homodyne RX issues, such as LO feed-thorough, dc offset, $1/f$ noise and second-order nonlinearity, which force all the active devices to be very large. Here, all the gain blocks are simple inverter-based g_m stages. All switches and capacitors, which are used in the filters, are amenable to technology scaling. This high-Q BPF has a superior image rejection as compared to the N-path filter. In mixer-based BPFs, such as the N-path filter, the rejection of the image components is ultimately limited by the mismatch between the LO clock of I and Q paths. On the other hand, there is no inherent limitation here on the level of image component rejection other than the NF degradation and power consumption of LO distribution.

The circuit of the on-chip complex notch filter is depicted in Fig. 5.15. The wanted signal at f_{IF} is downconverted by the mixers and filtered through the C-R filter, which acts as a HPF at dc. Then, the signal is upconverted to the IF frequency with the second mixer. Similar to the N-path filter, harmonic mixing might also happen in the

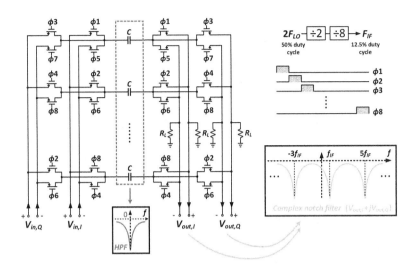

Figure 5.15 Circuit level of the complex notch filter centered at f_{IF}.

N-path notch filter. However, it is not an issue in this RX since the image components are already filtered out via the preceding complex wideband IQ charge-sharing filter. The 8-phase clock for the notch filter is provided by dividing the main LO by 2 and then further dividing it by 8, through the chain of ÷2 dividers, The ÷2 divider ensures that the 8-phase output clocks are nonoverlapped.

5.2.2   Measurement Results

The receiver (RX) chip is fabricated in 65 nm CMOS technology. The input signal lies in the range of 500 MHz to 1.2 GHz, corresponding to the IF frequency of 33.33 MHz to 80 MHz. The C_H and C_r capacitors are binary adjustable between 3.8–11 pF and 1.2–2 pF, respectively. The notch filter capacitance (C in Fig. 5.15) is 5 pF . The measured RX gain is 35 dB and the NF is 6.7 dB at the max gain. The in-band IIP3 is +10 dBm at the 25 dB gain with a two-tone test at +5 MHz and +10 MHz; it is 0 dBm at +1 MHz and +2 MHz. The measured RX transfer function at various LO frequencies is demonstrated in Fig. 5.16. The notch in the transfer function is due to the dc block capacitors in the feedforward path of the g_m stages shown earlier in Fig. 5.12. This further improves IM2 and clock feedthrough. The BW of the RX is 4.5 MHz. It can be seen that the rejection around the RF frequency is more than 10 dB. The images at $7f_{IF}$ and $9f_{IF}$ could theoretically be folded into the wanted signal in the complex notch filter. However, this is not an issue because these images are already rejected through 35 dB attenuation in the IQ charge-sharing BPF. Note that no preselect filters are used here. Therefore, any possible folded images from $7f_{IF}$ and $9f_{IF}$ are first attenuated by 53 dB (35 dB+18 dB). On the other hand, it should be possible to employ the high-Q N-path filter in the feedforward path to improve the filtering function after the IQ charge-sharing BPFs. The two "shoulders" around f_{RF}

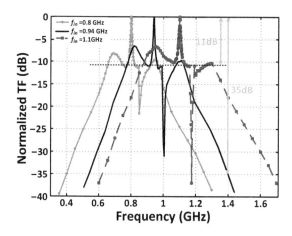

Figure 5.16 Measured transfer function of the receiver.

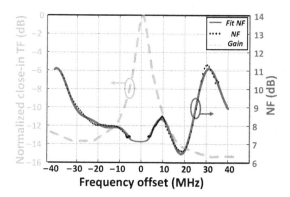

Figure 5.17 Measured transfer function and NF around desired RF frequency versus frequency offset.

in Fig. 5.16 are due to the transition from the filtering function of the sharp high-Q RF BPF to the IQ charge-sharing BPF. The measured close-in transfer function and NF are depicted in Fig. 5.17.

The Q-factor of the BPF is 208 and the total power consumption of the RX is 24.5 mA. The performance of the RX is summarized and compared in Table 5.2 with the only other published high-IF RX [8]. A clock generation circuit consumes 6 mA at 1.2 V. The active area of RX including the clock generation is 0.45 mm², as shown in Fig. 5.18. The presented high-IF RX with high-Q complex BPF offers superior filtering at RF frequencies in addition to the strong filtering of the image components, while achieving low power consumption in a very small chip area. It should be emphasized that the LNA was not implemented in this chip in order to better characterize the linearity and noise. Hence, the NF given in Table 5.2 is high, just as expected, due to the low gain of the RX front-end (i.e., g_m and mixer) stage, which is about 6 dB.

Table 5.2. Summary and comparison with state-of-the-art

	This work	[8]
CMOS technology	65 nm	65 nm
Active area	0.45 mm^2	0.76 mm^2
Power consumption	24.5 mA	21 mA
Rejection @ $f_{LO} - 7f_{IF}$, $f_{LO} + 9f_{IF}$	<-53 dB	<-18 dB
NF (dB)	7.5	2.8
IIP3 (dBm) @ 1M, 2M	0	–
IIP3 (dBm) @ 5M, 10M	+10	–
IIP3 (dBm) @ 10M, 20M	+2	–
BW (MHz)	4.5	4
RX frequency (GHz)	0.5–1.2	1.8–2.2

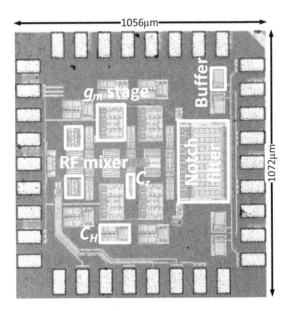

Figure 5.18 Chip micrograph.

5.3   DT Superheterodyne Receiver in 65 nm CMOS

This section presents a fully integrated DT superheterodyne receiver. The receiver uses the concepts and blocks introduced in Chapters 3 and 4. After discussing this structure and its DT model, the internal frequency translations as well as image rejection mechanisms are explained. Then, baseband signal processing using an ultralow

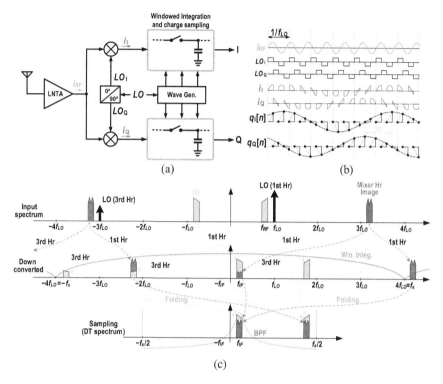

Figure 5.19 (a) Signal sampling after the mixer in a DT receiver with (b) its time-domain signal waveforms (c) Frequency translation considering CT mixing and then sampling. Although only the main and third harmonics of mixer are shown, it still make a rather complicated frequency translations.

power implementation is discussed. Afterward, the measurement results of the implemented test chip in CMOS 65 nm are presented and compared with other state-of-the-art receivers.

5.3.1   Structure

Figure 5.19(a) shows the simplified DT receiver. In this RX, signals at the mixer outputs are still continuous-time (CT). In addition to the main harmonic of the mixer clocks, $LO_{I/Q}$, they also possess odd harmonics (i.e. +3rd, −5th, +7th, etc.) due to their square-like waveforms (see Fig. 5.19(b)). These harmonics not only down-convert high-frequency images on top of the wanted signal, but also upconvert the input spectrum to high frequencies around the harmonics. In reality, the windowed current integration, sampling, and DT processing of samples happen in the subsequent switched-capacitor block. As mentioned earlier, the sampling folds the spectrum of the signal that is outside of the Nyquist range into the $-f_s/2$ to $+f_s/2$ range. Since both mixing and sampling processes translate frequencies with respect to the LO harmonics

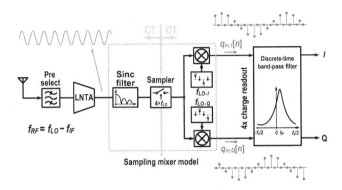

Figure 5.20 DT superheterodyne receiver model using 4× sampling.

and sampling rate, respectively, they make a rather complicated matrix to calculate the total frequency translations from RF input to the sampled output of the receiver (see Fig. 5.19(c)).

The top-level diagram, shown in Fig. 5.20, provides a straightforward yet accurate model for the DT receiver, illustrating its functionality and the scheme of frequency translations. Since the accumulated charge is read out by the switched-capacitor filter at the 4× rate, and also the states of mixer clocks are changing with the same rate (i.e., four times in each cycle), these operations are mutually commutative, so it would make no difference if we consider the WI and sampling executed ahead of the mixing. In this way, the rest of signal processing after the sampling is done in the discrete-time domain . Therefore, the *DT mixers* interpret their input signals as DT input sequences instead of the CT square waveform. Moreover, the outputs of DT mixers become sampled-charge data rather than the CT i_I and i_Q waveforms of Fig. 5.19(b). This model will be later used to calculate the gain and to illustrate the frequency translations of the DT receiver. Moreover, this model can also be used for other IF sample rates (e.g., $1 \times /2\times$) by placing a DT decimation block after the DT mixer in Fig. 5.20.

Discrete-time charge packets after windowed integration and sampling are described with the following equation [5, 11, 7, 46]:

$$q_{in}[n] = \int_{(n-1)T_s}^{nT_s} i_{RF} dt, \tag{5.3}$$

where i_{RF} is the result of a voltage at LNTA input (V_{RF}) multiplied by the transconductance g_m of LNTA. This WI creates a continuous-time sinc-type filter prior to the sampler in Fig. 5.20:

$$H_{WI}(f) = T_s \times \frac{\sin(\pi f T_s)}{\pi f T_s} = T_s \times \text{sinc}\left(\frac{f}{f_s}\right). \tag{5.4}$$

This filter has notches at multiples of f_s.

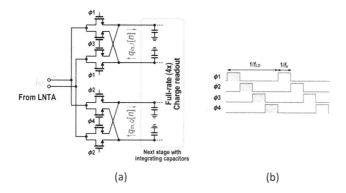

Figure 5.21 (a) Implementation of the sampling mixer in Fig. 5.20 with passive current commutating mixer and (b) driving clock waveforms.

5.3.2   Sampling Mixer

The clock sequences of DT mixers in Fig. 5.20 are repetitively $LO_I[n] = \{1, 0, -1, 0\}$ and $LO_Q[n] = \{0, -1, 0, 1\}$. They could be written as:

$$
\begin{cases}
LO_I[n] = \frac{1}{2}e^{j(\frac{\pi}{2}n)} + \frac{1}{2}e^{-j(\frac{\pi}{2}n)} \\
LO_Q[n] = \frac{1}{2}e^{j(\frac{\pi}{2}n+\frac{\pi}{2})} + \frac{1}{2}e^{-j(\frac{\pi}{2}n+\frac{\pi}{2})}.
\end{cases}
\tag{5.5}
$$

In the frequency domain, they have two tones at $\pm f_s/4$, which is $\pm f_{LO}$. From (5.5), the downconversion (as well as upconversion) gain of each DT mixer becomes $A_{\mathrm{mix},I/Q} = 1/2$. However, the output of the mixer in the Q path has a 90° phase shift with respect to the I path.

Implementation of the sampling mixer circuitry is depicted in Fig. 5.21(a). It consists of two passive double-balanced mixers for I and Q paths. Each mixer commutates its input current provided by the LNTA. While input of the RF mixer is actually a CT current, their outputs are interpreted as DT charge packets, resulting from the windowed integration. Each new coming phase generates a new charge sample. The next block, DT BPF, should also read the charges at the 4× rate to maintain the sample rate.

Mixer switches are implemented by NMOS-only. As depicted in Fig. 5.21(b), nonoverlapped clocks with a 25% duty cycle are used to drive the switches. This avoids charge cross-talk between I/Q paths, which would compromise functionality of the subsequent block. Moreover, driving switches with the 25% duty cycle improves IIP3, noise and gain of the mixer, as compared to the 50% duty cycle [22]. In our implementation, the sampling mixer works up to 12 GS/s, limited only by the digital clock generator circuit.

5.3.3   Receiver Chain

Receiver chain from RF input to the end of IF strip is shown in Fig. 5.22. Four charge-sharing BPFs are cascaded to increase the rejection of out-of-band blockers

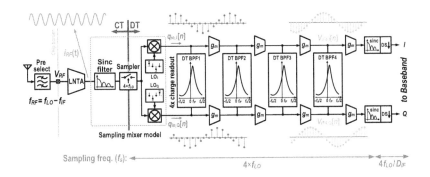

Figure 5.22 DT superheterodyne receiver using $4\times$ sampling.

or interferers. In addition to providing isolation, each of the g_m-cells between the BPFs serves as a DT gain stage, providing 0–6 dB signal gain. In this way, the noise of each subsequent block gets less important. All the CS-BPFs operate with the same $4\times$ sampling rate. Until this stage, the receiver has had enough out-of-band filtering. Therefore, to save power consumption, the signal can be decimated, and the baseband circuit is clocked at the lower rate. To do so, an additional g_m-cell stage is used at the end of IF strip that again converts voltage to charge. This creates a sinc antialiasing filter before the decimation.

5.3.4   Frequency Translation

The whole process of frequency translations that happen in the presented HIF DT receiver is depicted in Fig. 5.23. As the continuous-time input signal enters the receiver, whose model is shown in Fig. 5.22, it is filtered by the CT sinc filter described in (5.4). Images are then created due to sampling, as indicated in Fig. 5.23(a). In this example, we choose RF signal at $f_{LO} - f_{IF}$ such that sampling images are at $-f_{LO} + f_{IF} + k(4 f_{LO})$ and $f_{LO} - f_{IF} + k(4 f_{LO})$ for $k = 1, 2, 3, \ldots$. From (5.4), sinc filter attenuation of the first two images ($k = 1$) near third and fifth $_{LO}$ harmonics are 9.5 dB and 14 dB, respectively. These levels are the same as the image attenuation of a CT 4-phase mixer. Although these numbers alone might not be adequate, the images are further attenuated in this receiver by the LNTA and a preselect filter. After sampling, the DT input spectrum is now spread from $-f_s/2$ to $+f_s/2$, where f_s is $4 f_{LO}$.

Figure 5.23(b) shows the wanted RF signal and the important images of the receiver. Complex LO signal defined in (5.5) is also displayed as a black single tone. After mixing the entire signal spectrum with this tone, the negative side is downconverted to around dc, while the positive side is upconverted to close to $\pm f_s/2$ (see Fig. 5.23(c)). At this point, the wanted signal is located at $+f_{IF}$ while its IF image at $-f_{IF}$. Because of the quadrature operation of the mixer, these two signals are not aliased, and the image can be filtered at IF and later at baseband (BB).

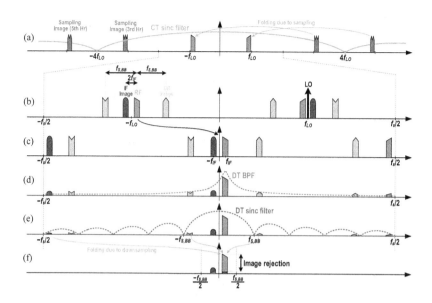

Figure 5.23 Frequency translations in the DT receiver: (a) images caused by sampling of CT signal, (b) input spectrum after the sampling, (c) downconverted spectrum after the DT mixer, (d) signals after IF filter stages; and decimation by (e) applying an antialiasing filter before (f) baseband downsampling

The spectrum of Fig. 5.23(c) is then filtered by the complex DT BPFs in the IF strip (see Fig. 5.23(d)). The wanted signal is amplified while the rest are attenuated. At this point, out-of-band images and blockers are attenuated enough such that the signal of interest can be decimated to a lower baseband sample rate, $f_{s,BB}$. This leads to power consumption reduction for the remainder of the processing blocks.

The decimation process is being protected by a DT sinc antialiasing filter that is simply achieved by adding up DIF samples (a.k.a., moving average filter), where DIF is a positive integer. Therefore, the images are further filtered out (Fig. 5.23(e)) before downsampling and aliasing over the wanted signal (Fig. 5.23(f)). The new sample rate becomes $f_{s,BB} = f_s/DIF$. Transfer function of the moving average (MA) filter is

$$H_{\text{MA,IF}}(f) = D_{IF} \times \text{sinc}\left(\frac{f}{f_s}\right). \tag{5.6}$$

A small resulting attenuation of the wanted signal at f_{IF} is neglected in the rest of the text. Implementation of the decimation is trivially achieved by lowering the readout rate of the block succeeding the g_m-cell. In this way, after several samples are accumulated, they are processed once (temporal decimation [68]).

Considering the frequency translations in Fig. 5.23 and the receiver model shown in Fig. 5.22, we are now able to calculate the gain of signals at different frequencies from the LNTA RF input to the IF strip output. The wanted RF signal experiences the LNTA gain, windowed integration sampling, DT mixer downconversion gain, passband gain of BPFs, and g_m-cells. Using (4.5), (5.4), and (5.5), we have:

$$G_{\text{wanted}} = \frac{V_{IF4,I/Q}}{V_{RF}} = [g_{m_{\text{LNTA}}} H_{WI}(f_{LO} - f_{IF}) A_{mix} A_{BPF})] \times [A_{g_{m_{IF}}} \cdot A_{BPF}]^3$$

$$\approx \left[g_{m_{\text{LNTA}}} \text{sinc}\left(\frac{1}{4}\right) \times \frac{1}{2} \times \frac{1}{f_s C_R} \right] \times \left[\frac{g_{m_{IF}}}{C_R f_s} \right]^3, \tag{5.7}$$

where $(A_{g_{m_{IF}}} A_{BPF})$ is defined as:

$$A_{g_{m_{IF}}} A_{BPF} = g_m T_s \times \frac{1}{C_R} = g_m \frac{1}{C_R f_s}. \tag{5.8}$$

In (5.7), the two terms are the gain of the stages until the first BPF, and the second to fourth BPFs, respectively. In the approximation, $f_{IF} \ll f_{LO}$ is considered. The closest image that could fold onto the wanted RF signal is the IF image at $f_{LO} + f_{IF}$. As shown in Fig. 5.23(e), part of the IF image energy after mixing and attenuation resides at $-f_s/2 + f_{IF}$. This signal is folded over the wanted signal after downsampling, assuming an even decimation factor, DIF. By neglecting a small gain difference due to the CT sinc filter (HWI), IF image rejection is calculated by adding the attenuations of the BPFs and DT moving average filter. Using (4.3), gain of BPF at $-f_s/2 + f_{IF}$ can be very well approximated by $1/(2C_H)$ for C_R/C_H. Therefore, the attenuation of BPF with respect to its passband gain in (4.5) is $2C_H/C_R$, for each of the four stages. Therefore, by adding the effect of the moving average filter from (5.6), IF image rejection ratio (IMRR) is

$$\text{IMRR}_{IF} = \frac{\left(\frac{2C_H}{C_R}\right)^4}{\text{sinc}\left(\frac{-f_s/2 + f_{IF}}{f_{s,BB}}\right)} \tag{5.9}$$

Considering $f_{IF} = f_{LO}/16$ in our implementation, (4.4) suggests C_H/C_R to be about 10. Consequently, with the DIF decimation factor of 16, the total IMRR_{IF} reaches more than 135 dB. Note that this value assumes perfectly quadrature LO signals driving the mixers without any mismatch. However, the quadrature inaccuracy of the practical LO signals also aliases a tiny part of IF image right after the mixers, from $f_{LO} + f_{IF}$ to $+f_{IF}$ in Fig. 5.23(b) and (c). The latter effect is predominant and limits IMRR_{IF} to 40–80 dB, depending on the quadrature accuracy, layout, and mixer mismatch.

The second important class of images are baseband downsampling images. Translated back to the RF input, they are located at $f_{RF} \pm k f_{s,BB}$. The first two of them (for $k = 1$) are shown in Fig. 5.23(b). After mixing down (Fig. 5.23(c)) and passing through the BPFs (Fig. 5.23(d)), they are attenuated by the DT moving average filter (Fig. 5.23(e)), and then folded over the wanted signal via downsampling (Fig. 5.23(f)). By means of (4.3), the exact attenuation of BPF can be calculated. As a first-order approximation of (4.3) for mid-range frequencies ($f_{IF} \ll f \ll f_s/2$), Bode plot of a first-order low-pass filter with a 3 dB bandwidth of f_{IF} is being considered that is shifted to be centered at f_{IF}. Therefore, BPF rejection at s,BB offset from the passband is approximated as:

$$R_{BPF}(f) \approx \left. \frac{f - f_{IF}}{f_{IF}} \right|_{f=f_{IF}+f_{s,BB}} = \frac{f_{s,BB}}{f_{IF}}. \tag{5.10}$$

Both sampling images are attenuated by the same amount, due to the symmetry around f_{IF}. The higher $f_{s,BB}$, the higher the attenuation. Then, the images are attenuated by the moving average filter in (5.6). A higher $f_{s,BB}$, makes the images relatively closer to the notches of the sinc filter and improves the attenuation. By adding up all these attenuations, the baseband downsampling IMRR becomes:

$$\text{IMRR}_{\text{BB}} = \left(\frac{f_{s,BB}}{f_{IF}} \right)^4 \Big/ \text{sinc}\left(\frac{f_{IF} \pm f_{s,BB}}{f_{s,BB}} \right). \tag{5.11}$$

By choosing a proper number of BPF stages and a decimation factor to set $f_{s,BB}$, a desired IMRR can be achieved.

5.3.5 Selection of IF Frequency

Based on (5.11), if the ratio of $f_{s,BB}/f_{IF}$ is fixed, changing IF would not have any major impact on the BB IMRR. If improving this rejection is desired, $f_{s,BB}$ has to be increased. In case ADCs are used at the end of IF chain, higher $f_{s,BB}$ means a higher ADC sample rate. On the other hand, if a fixed ratio is considered, lowering f_{IF} results in a narrower bandwidth of BPF's, and hence, higher linearity (IIP2 and IIP3) at a fixed offset frequency. However, f_{IF} should always be higher than the IM2 product and the flicker noise corner. In this work, for the sake of simplicity, a sliding IF approach with $f_{IF} = f_{LO}/16$ is used. With $f_{s,BB} = 4 f_{IF}$ used in our analog baseband implementation, theoretical BB IMRRs could reach 59 and 63 dB for the images at $f_{RF} + f_{s,BB}$ and $f_{RF} - f_{s,BB}$, respectively. In transistor-level simulations, 46 dB and 51 dB rejections are obtained, respectively. The shortfall is due to lowering of the quality factor of BPFs by the output resistance of IF g_m-cells

5.3.6 Baseband Signal Processing

The signal at the end of IF strip can be directly sampled and digitized using Nyquist-rate or band-pass ADC [8]. Afterward, baseband signal processing, including IF mixing and channel select filtering, can be done entirely in digital domain. Among various ADC architectures, especially SAR ADC is attractive for this receiver, due to its process scalability and compatibility with digital CMOS technology. However, this approach might not be always the most power efficient because of the high sample rate and high dynamic range requirement imposed on the ADC and subsequent digital signal processing. The alternative approach chosen in our 65 nm CMOS implementation is signal processing through the DT analog baseband, as shown in Fig. 5.24. The main goal of this part of the receiver is, by means of filtering and decimation, to reduce the required ADC sample rate and dynamic range. The introduced DT baseband consumes only a few miliwatts, while significantly reducing power consumption of the ADC and the digital baseband.

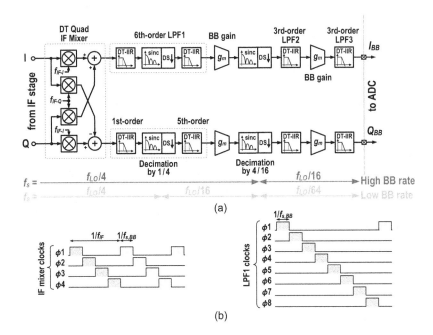

Figure 5.24 (a) Baseband DT signal processing of the receiver. (b) Required clock waveforms for the IF mixer and LPF1.

DT Analog Baseband Signal Processing

The first stage of the analog baseband circuitry is a quadrature DT IF mixer. A set of four mixers downconvert the signal from IF to zero. Implementation of each mixer is similar to the passive RF mixer as shown in Fig. 5.21. The baseband sample rate (fs,BB) is chosen 4fIF to simplify the generation of clocks needed for the IF mixer. As shown in Fig. 5.24, the IF clock waveforms are very similar to those in the RF mixer, but the period is $1/f_{IF}$. If a higher attenuation of the sampling images would be required (e.g., for a SAW-less application), a higher baseband sample rate combined with harmonic rejection mixing could be implemented. The IF mixer is the only circuitry in this receiver that limits the overall IIP2, even though it is still extremely high. Since the IF mixer is clocked at a much lower rate than the RF mixer, its IIP2 is substantially better [28]. Moreover, the IF filtering considerably improves its IIP2 referred back at the antenna (each 1 dB of a blocker filtering at preceding stages improves IF mixer IIP2 by 2 dB).

The second stage of the analog baseband is a DT sixth-order low-pass filter (hereafter, IIR6), based on the work in [5]. This narrow-bandwidth filter selects the wanted signal and attenuates the rest of in-band and out-of-band interferers and blockers. Thanks to its high filtering order, IIP3 requirements of all the following stages are relaxed. Sample rate of this filter is also the same as $f_{s,BB}$. Figure 5.25 shows a switch-level implementation of this filter. C_{H1} at the input port accumulates the input charge. Through a prearranged switching sequence, each of the C_S capacitors rotates the partial charge of C_{H1} to other history capacitors, C_{H2-6}, and then gets reset.

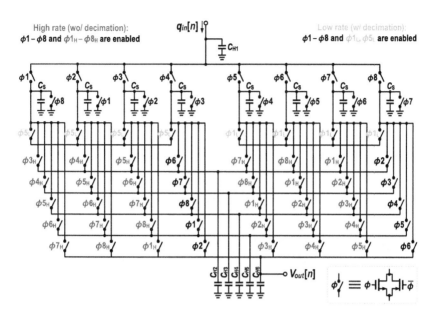

Figure 5.25 Implementation of the DT sixth-order IIR low-pass filter with selectable decimation by 4.

Each charge-sharing operation within the cycle adds one order of filtering. Using eight sampling capacitors, each with a delay of one phase, increases the filter's sampling rate eight times while using the same clock signals (parallelized operation). More details and analysis of this filter are described in [5]. In the normal high sample rate mode, ϕi and ϕi_H, $i = 1, \ldots, 8$ are clocked and the filter works as described. This mode is used for high bandwidth signals up to about 30 MHz (e.g., for the LTE standard). For narrowband signals (e.g., 200 kHz in GSM standard), the sample rate of $4 f_{IF}$ (several 100 s of MS/s) would be much higher than necessary. Further decimation should therefore be done to save power consumption. In this low sampling rate mode, only switches with clock phases as ϕi and ϕi_L, $i = 1, \ldots, 8$ are engaged and switches with clock phases as ϕi_H, $i = 1, \ldots, 8$ are disengaged to save power. After a set of four succeeding C_S's are charge-shared with C_{H1}, they are shorted together to make a spatial decimation by 4 [68]. Charge sharing of the four C_S's makes a 4-tap moving average (sinc antialiasing filter) prior to the subsequent decimation. Then one of them continues charge sharing with C_{H2-6}. This also reduces the required C_H value to support the narrow bandwidth. In this mode, the input sample rate of this filter is $4 f_{IF}$ while its output rate is reduced to f_{IF}. Clock waveforms required for driving this filter are shown in Fig. 5.24.

The receiver path up to the end of IIR6 already enjoys high gain and high order of filtering. Hence, noise and IIP3 of the remaining stages would be less of a concern. Due to this reason, the following stages can be implemented at an ultralow power consumption. After IIR6, two filter stages are cascaded with two g_m-cells. Figure 5.26 shows the implementation of the baseband g_m-cells in which a fully differential

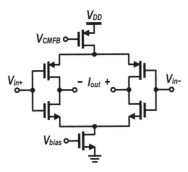

Figure 5.26 Implementation of baseband g_m-cells.

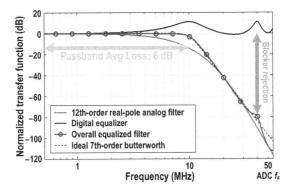

Figure 5.27 Digital equalization of 12th-order real-pole transfer function to better than a 7th-order Butterworth filter. The ADC and digital equalizer are clocked at 50 MHz.

inverter-based structure is used. Note that the baseband g_m-cells are dc-coupled to pass the whole signal spectrum. Although putting tail current sources in the g_m-cell reduces IIP3, the cell still features good enough linearity (IIP3 = 0 dBm) considering the filtering beforehand. Both stages of the third-order IIR LPFs (*IIR3*) are identical and use the same structure as in Fig. 5.25, though without spatial decimation [5]. To further save power in these filters, their clocks are reduced by $4\times$ with respect to the IIR6 output rate. This creates a temporal decimation after the first baseband g_m-cell. The two baseband gain stages together with LPF2 and LPF3 in I and Q paths totally consume only 100–700 µW, depending on the gain and bandwidth setting.

At the bottom of Fig. 5.24 (a), the sample rate of each block from IF to the end of the baseband is displayed. Due to the high total order of filtering (up to the 12th-order in this design), ADC sample rate could be reduced to below the receiver output sample rate without any other antialiasing filter.

5.3.7   Digital Equalization

The analog baseband part of the receiver features the total of 12th-order of filtering. However, despite this very high order, only real poles are used, and so this filter cannot be directly compared with complex-pole filter types (e.g., *Butterworth* or *Chebyshev*) of the same order. A high-order real-pole filter provides a gradual and smooth transition between its passband and its sharp out-of-band roll-off (Fig. 5.27). Therefore, [5] has presented using a low-power digital equalizer after the ADC to map the real-pole transfer function to a sharp complex-pole filter, but with a lower order. In this way, the passband of the filter experiences a small average loss. The lower order of the mapped filter, the lower the passband loss [5]. For example, in Fig. 5.27, the 12th-order real-pole filter is considered to be mapped to a better than seventh-order Butterworth filter (that should be enough for BB filtering of most wireless standards). As has been calculated, the passband loss is 6 dB that can be compensated by the preceding gain stages or 1 additional ENOB in the ADC [5]. Taking into consideration the complete system-level view of the receiver, the introduced baseband processing consumes several times less power (total I/Q baseband: 2.3 mW for 1.96 GHz RF input) than the conventional CT or active switched-capacitor approaches [101, 103], while providing a much lower NF and a very high linearity [5].

5.3.8   Clock Waveform Generator

All the required clocks for the RF mixer (Fig. 5.21) and IF BPF stages (Fig. 5.23) are identical 25% duty cycle clock waveforms at the LO frequency. First, an external clock at 2fLO is fed in, then divided by 2 to generate four quadrature 50% clocks (LO1-4 in Fig. 5.28 (a)). The divider consists of two latches arranged in a loop with a crossed feedback, providing an additional 180° phase shift. A similar approach is also used in [8]. As shown in Fig. 5.28(b) and (c), two clock-gated (tristate) inverters with weak back-to-back inverters are used as a dynamic latch. This promotes a very high speed of operation at low power. Then, NAND gates are used to make the four 25% clocks (ϕ_{1-4}). Skewed NMOS and PMOS transistors in the NAND gates and their following buffers are used to ensure nonoverlapped clocks. Required clocks for baseband are generated in a similar way by using standard cells, whose reference clock is provided

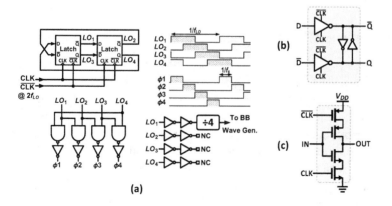

Figure 5.28 (a) RF and IF waveform generator. (b) Dynamic latch using (c) gated inverter.

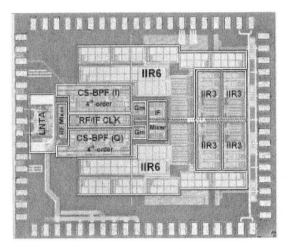

Figure 5.29 Receiver's chip micrograph occupying $1.9 \times 2.4\,\mathrm{mm}^2$.

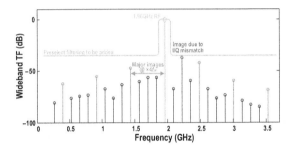

Figure 5.30 Measured wideband transfer function.

via a divide-by-4 of LO_1 clock. To keep the phases balanced, two levels of dummy buffers are used for LO_{2-4}. Transistors in the clock generator circuit are sized based on the phase noise requirement and load capacitance.

5.3.9   Measurement Results

The receiver is implemented in standard TSMC 1P7M 65 nm CMOS and occupies an active area of $1.1\,\mathrm{mm}^2$ (Fig. 5.29). It consists mostly of MOS switches, capacitors, and inverter-based g_m-cells, making it process scalable and amenable to digital nanoscale CMOS. Majority of the chip area is occupied by capacitors used for baseband filtering that supports cutoff frequencies down to 100 kHz. Most of the capacitors are of metal-oxide-metal (MOM) type implemented differentially. Therefore, the chip area scales very well with the CMOS technology advancements.

Measured wideband transfer function of the complete receiver is plotted in Fig. 5.30. There are only a discrete number of frequency points that can fold into the received band of interest. As analyzed in Section 5.3.4, the major images are

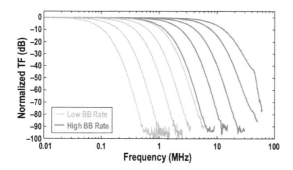

Figure 5.31 Measured close-in transfer function of the receiver.

located at multiples of $4f_{IF}$ away from f_{RF}. The first two major images are rejected by 42 and 46 dB, close to the simulated values. The remainder of images at f_{IF} multiples (in black) are much smaller and are caused by the baseband decimations. The exception is the image of 37 dB rejection that is due to an uncalibrated I/Q clock mismatch. There, unaccounted parasitics on the mixer clock lines make the I/Q unbalanced. Based on simulations, a phase mismatch of about 1° could lead to the measured degraded rejection. A more careful layout design solves this in future designs (in [15], a 65 dB I/Q matching is achieved). Including the preselect filter with a moderate OOB rejection of 35 dB that precedes the receiver, the total image rejection easily improves to better than 72 dB.

Measured close-in transfer functions of the receiver for different programmed bandwidths at low or high baseband rates are shown in Fig. 5.31. While the IF filters are designed mostly to reject out-of-band blockers or interferers, the close-in TF is created by the baseband filtering to select the desired channel. After converting to digital by the ADC, a digital equalizer will flatten the passband of the transfer function, per the mixed-signal scheme of Fig. 5.27. As a whole, the RF bandwidth of the receiver is programmable from 200 kHz to 30 MHz.

Figure 5.32(a) shows the measured uncalibrated IIP2 and IIP3 of the receiver at the medium gain setting in which the receiver meets the sensitivity specification in the presence of blockers. In-band IIP3 is measured at −5 dBm, which is mainly limited by the linearity of IF g_m-cells. While the high-IF front-end has infinite IIP2, the IF mixer limits the receiver's IIP2. The IF filters in this receiver attenuate the blockers and, therefore, out-of-band IIP2 increases rapidly at higher frequency offsets (Fig. 5.27(b)), from +47 dBm in-band to +93 dBm at 120 MHz offset, all uncalibrated. Although this receiver does not claim to be SAW-less, the superheterodyne architecture appears to be the path to reach such SAW-less operation that could meet the most stringent IIP2 requirements, even in the FDD mode and without any calibration.

Plotted in Fig. 5.33, the noise figure of the complete receiver for different bands from 400 MHz to 2.9 GHz ranges between 2.9–4.0 dB. For each RF frequency, NF is the average value over the BW (anyway it stays constant over up to three times BW). At higher frequencies, the duty cycle of ϕ_{1-4} RF clocks is reduced because of a

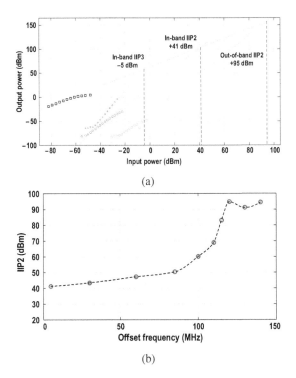

(a)

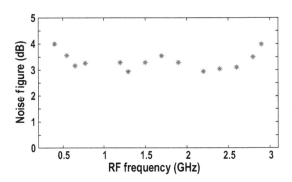

(b)

Figure 5.32 (a) Measured IIP2 and IIP3 (b)Measured IIP2 versus offset frequency.

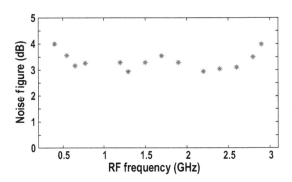

Figure 5.33 Measured noise figure of the complete receiver versus RF frequencies.

limited rise/fall times. As a consequence, the gain of RF mixer reduces, which directly degrades the RX noise figure.

Table 5.3 summarizes the measured performance of the receiver and compares it with the published state-of-the-art. The analog part of the receiver consumes 43 mW in total for the high-gain setting. The clock waveform generator consumes 5–36 mW that linearly scales with f_{LO}.

Figure 5.34 shows the power consumption budgeting of different blocks. Full chain of the receiver has a maximum gain of 83 dB.

Table 5.3. Performance summary and comparison with state-of-the-art receivers

	This work	[6] RFIC'13	[8] JSSC'11	[91] JSSC'10	[11] JSSC'06	[13] JSSC'14	[19] JSSC'14	[22] JSSC'9
Technology	65 nm	65 nm	65 nm	90 nm	90 nm	65 nm	28 nm	90 nm
Architecture	Superhet.	Superhet.	Superhet.	Zero-IF	Zero-IF	Zero-IF	Zero-IF	Zero-IF
Description	Full DT	DT/N-path	N-path	Full DT	CT/DT	DT/CT	Full CT	Full CT
Analog BB/Order	Yes/7th[a]	No	No	Yes/2nd[b]	Yes/3rd[b]	Yes/2nd[b]	Yes/2nd[c]	Yes/2nd[c]
RF frequency (GHz)	0.4~2.9	0.5~1.2	1.8~2.2	0.5~3.8	0.8~6	0.5~3	0.4~6	0.8~2.2
Supply voltage (V)	1.2/2	1.2	1.2/2.5	1.2	1.0/2.5	1.2/2.5	0.9	1.5
Power (mW)[d]	48~79	25.4	39	67~115	45.5~65.5	210~540[e]	35~40	19.5~22.6
NF (dB)	2.9~4	7.5	2.8	5.3~6	5~5.5	5.5~8.8	1.8~3.1	2.2~3.2
Max. gain (dB)	83	35	55	58/64	>47	35	70	61.5
In-band IIP3 (dBm)	−5	+10	−8.5	+1/+2.5	−35	>−12.5	+4	–
Out-of-band IIP3 (dBm)	+93	–	–	+38~+52	+60	>+46	+80	+90
calibration	No	–	–	No	No	No	Yes	Yes
Channel BW[f] (MHz)	0.2~30	4.5	4	0.2~20	0.2~20	26	1~100	0.2~3.8
Area	1.1	0.45	0.76	0.5	3.8[g]	1.85[e]	0.6	–

[a] 12th-order real pole mapped to a 7th-order Butterworth.   [b] Real-pole.   [c] Biquad.   [d] At highest gain setting.   [e] Synthesizer and bias are excluded.
[f] Two times BB bandwidth.   [g] Including synthesizer.

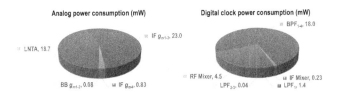

Figure 5.34 Power consumption budget of various blocks for maximum gain setting at 1.96 GHz RF input.

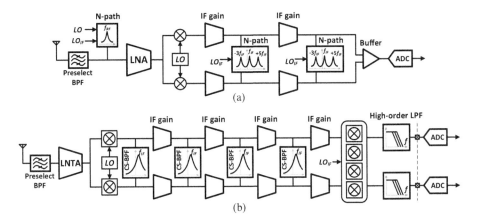

Figure 5.35 State-of-the-art superheterodyne receivers: (a) with N-path filter. (b) With complex BP IIR filter.

5.4   SAW-less DT Superheterodyne Receiver in 28 nm CMOS

The integration problem of high-IF BPF was addressed in [8] (see Fig. 5.35(a)) utilizing an N-path filtering technique [115, 116, 111, 117, 119, 118]; and in [45, 1] (see Fig. 5.35(b)), [6] using a discrete-time (DT) quadrature charge-sharing (CS) BPF [128, 122]. The N-path filter cannot reject images defined as blockers/interferers at harmonics of the IF frequency because it *inherently* features replicas there [8]. On the contrary, the transfer function of the DT CS-BPF has only one peak in the entire sampling frequency domain of $-f_s/2$ to $f_s/2$, which makes it a proper candidate as an integrated BPF for superheterodyne receivers [128]. The center frequency and bandwidth of the full-rate DT CS-BPF in [45, 6] are precisely controlled via f_s and the capacitor ratios. In addition, that filter comprises only transistors as switches and capacitors, which occupy a small area and follow the process scaling very well. Unfortunately, CS-BPF in [45, 6] has insufficient blocker rejection to support the SAW-less operation.

We introduce the superheterodyne architecture shown in Fig. 5.36 that utilizes a novel charge-sharing BPF based on an M/N-phase signaling and an extra pole to improve filtering. Combined with the presented highly linear wideband LNTA and cascaded harmonic rejection (HR) stages, the first-ever SAW-less high-IF

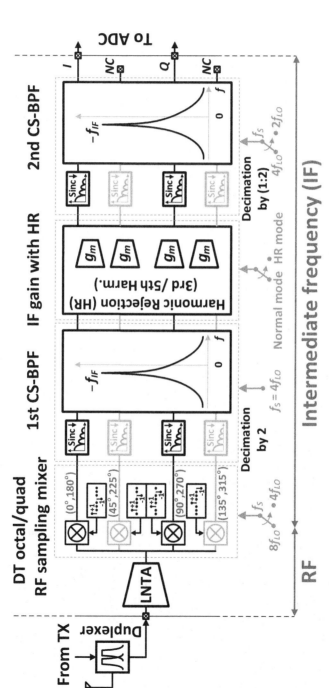

Figure 5.36 Superheterodyne receiver architecture including two stages of CS-BPF filtering and three stages of harmonic rejection.

(superheterodyne) RX is thus demonstrated. By exploiting two stages of the M/N-phase CS-BPF, the desired signal is amplified while the images and in-band/out-of-band blockers are progressively filtered out thoroughout the receiver chain.

As stated before, this architecture has several key advantages compared to state-of-the-art LIF RXs. First, since its IF is high, the issues associated with LIF RXs are eliminated, especially IIP2 and the need for dc offset calibration. Also, $1/f$ noise is not a concern anymore, so the active IF amplifiers use minimum length transistors. Second, two stages of DT CS-BPF consist of only capacitors as information charge storage devices, and transistors as switches. All of this makes the structure *fully* compatible with the technology scaling. Moreover, the presented RX offers the same level of monolithic integration as LIF RXs *without* using any calibration. Furthermore, this RX exhibits clear advantages over the traditional superheterodyne RXs. First, it includes two stages of integrated blocker-tolerant complex image-reject CS-BPFs and three stages of harmonic rejection circuitry. Second, since the center frequency (i.e., coinciding with the chosen IF) of the M/N-phase DT CS-BPF is well controlled by the clock frequency and ratio of capacitors, the IF frequency could be changed, thus avoiding RX desensitization in face of extremely large blockers. Finally, the second mixer and baseband filters have moved to the digital domain after the ADC (external in this work), hence they are ideal.

5.4.1   New SAW-less Superheterodyne Receiver

Digital circuits benefit from process scaling in both speed and energy due to, respectively, the increase in transistor transit frequency, f_T, and lowering of its dimensions with every finer process technology node. However, the analog/RF circuitry is getting worse, except for LNAs,[1] because the threshold voltage, V_{th}, remains almost constant while the supply voltage, V_{DD}, decreases. Moreover, the intrinsic gain and signal swing are reduced. All of those make analog/RF circuitry not amenable to CMOS scaling [129–132, 3].

On the other hand, the DT approach is fundamentally based on building blocks that scale very well: transistors acting as switches, switched capacitors, inverter-based g_m-cells, and digital clock generation circuitry. Hence, the RF performance improves with newer CMOS technology [9, 45]. These reasons motivate us to exploit the DT approach in this SAW-less superheterodyne RX shown in Fig. 5.36.

The input voltage at the antenna is converted to current by LNTA and down-converted to high-IF by DT sampling RF mixer, as shown in Fig. 5.36. The octal (i.e., 8-phase) mixer can be reconfigured to operate in the quadrature (i.e., 4-phase) mode if the detected reception conditions are not demanding. After the mixer, the sampled downconverted signal is fed to the DT CS-BPF to attenuate images and out-of-band blockers. To reduce the power consumption of the first CS-BPF even further, the decimation by 2 can be performed by integrating two samples, thus giving rise to the antialiasing sinc-type transfer function. In addition to all advantages of the

---

[1] LNA noise figure improves when f_T increases.

two-stage CS-BPF, each of them provides an intrinsic third or fifth harmonic rejection that can be further improved by turning on the additional HR block. The second CS-BPF is cascaded via inverter-based g_m-cells, providing flicker noise-free gain. The sufficient front-end filtering provided by the two-stage CS-BPF (unlike in [45]) allows to directly digitize the IF signal using a low-power ADC, and move the *second mixer* and baseband filtering into the digital domain. As calculated, a 10-bit 400 MS/s ADC should be sufficient after the two stages of CS-BPF filtering, while consuming less than 2 mW with state-of-the-art SAR ADC [133]. Also, it should be mentioned that the IIP2 generated by ADC is not a concern, because the ADC's IM2 component is at dc and the desired signal is at IF. The only possible limitation on the IIP2 in the introduced receiver is the quantization noise of the second digital mixer, but it can be arbitrarily reduced by increasing its word length.

5.4.2 DT Charge-Sharing Band-Pass Filter (CS-BPF)

The DT CS-BPF exhibits clear advantages over the traditional types of filters, such as active-RC, N-path, g_m-C, and biquad. The active-RC and g_m-C filters are substantially noisier due to the noise contributions from opamp and g_m components. Those components also generate flicker noise, so to suppress it, their area needs to be very large. Furthermore, typical IF and BB filters need to be reconfigurable in which the required bandwidth scales over a decade. Since the bandwidth in active filters is determined by the RC or C/g_m time constant, the capacitors should be up to 50% larger to compensate for RC and g_m-C mismatches. This contributes to their area disadvantage. As far as the N-path filters are concerned, they suffer from replicas at the harmonics of their mixer switching frequency, while CS-BPF has only one peak in the entire sampling frequency. In addition, in the traditional N-path filter, the stop-band rejection is severely limited by the switch on-resistance.

Further in this section, the causes leading to the creation and evolution of CS-BPF are detailed in Section 5.4.3 followed by a description of the introduced 8/8-phase and 8/16-phase CS-BPFs in Sections 5.4.4 and 5.4.5, respectively. After detailed comparison of different kinds of M/N-phase CS-BPFs, the general z-domain transfer function of M/N-phase CS-BPF is derived.

5.4.3 Conventional Quadrature CS-BPF

Figure 5.37(a) shows the well-known DT IIR LPF [134]. The input current i, generated by a g_m-cell, is integrated on the history, C_H, and rotating, C_R, capacitors as the input charge packet $q_0 = \int_{(n-1)T_s}^{nT_s} i \, dt$ during ϕ_1 over a time window T_s. At ϕ_1 going inactive, C_R samples a portion of the total "history" charge. As a result, the DT circuit illustrated in Fig. 5.37(a) has a first-order IIR characteristic, with C_R acting as a lossy component (termed "switched-capacitor resistor"). The order of Fig. 5.37(a) DT IIR filter can be further increased to second or fourth, as shown in Fig. 5.37(b) and Fig. 5.37(c), respectively; or indefinitely beyond, as demonstrated

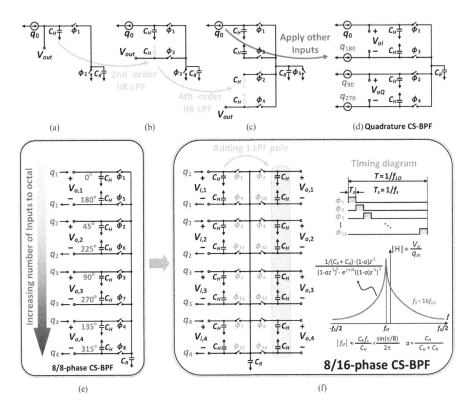

Figure 5.37 Evolution toward (f) M/N-phase CS-BPF ($M = 8, N = 16$), starting from (a) the simplest first order IIR LPF, then through (b) second order IIR LPF, through (c) fourth order IIR LPF, through (d) conventional CS-BPF, and finally through (e) 8/8-phase CS-BPF.

in [5]. The conventional quadrature CS-BPF with a single real-valued output can be synthesized from the fourth-order DT IIR filter by applying input charge packets q_0, q_{90}, q_{180}, and q_{270} with a multiple of $90°$ phase shifts, as shown in Fig. 5.37(d) [128]. By defining the complex-valued input constructed from two differential signals having the quadrature relationship, $q_I = q_0 - q_{180}$ and $q_Q = q_{90} - q_{270}$, the complex transfer function of a conventional quadrature CS-BPF is derived as:

$$H(z) = \frac{V_{oI}(z) + jV_{oQ}(z)}{q_I(z) + jq_Q(z)} = \frac{k}{1 - (a + j \cdot (1 - a))z^{-1}}, \tag{5.12}$$

where

$$k = 1/(C_H + C_R), \tag{5.13}$$

$$a = C_H/(C_H + C_R). \tag{5.14}$$

This transfer function has a first-order complex BPF characteristic with its peak located at:

$$f_{IF} = \frac{f_s}{2\pi} \arctan\left(\frac{1 - a}{a}\right). \tag{5.15}$$

The filter comprises only capacitors and switching transistors. Its center frequency f_{IF} only depends on the sampling frequency f_s and capacitor ratios. Hence, it is fully amenable to process scaling.

5.4.4 8/8-Phase CS-BPF

The filtering characteristic and tolerance to out-of-band blockers of the conventional quadrature CS-BPF can be *significantly* enhanced by increasing the number of inputs, corresponding history capacitors, and digital clock phases to 8 (i.e., octal) or more. As an example of such a filter, the schematic of a 8/8-phase CS-BPF is introduced in Fig. 5.37(e), where it features 8 inputs/outputs, 8 history capacitors, and 8 digital clock phases. The inputs, which are generated by the DT mixer for the first filter, are differential integrated charge packets q_1, q_2, q_3, q_4 that are phase shifted by $0, 45°, 90°$, $135°$. As in the traditional CS-BPF, C_R shares the charge between various C_H's. By defining the complex output voltage as

$$V_{oC} = V_{o,1} + e^{j\pi/4}V_{o,2} + e^{j\pi/2}V_{o,3} + e^{j3\pi/4}V_{o,4}, \tag{5.16}$$

and complex input charge as

$$q_{iC} = q_1 + e^{j\pi/4}q_2 + e^{j\pi/2}q_3 + e^{j3\pi/4}q_4, \tag{5.17}$$

and following the same approach as presented in [128], we find the complex transfer function of the 8/8-phase CS-BPF, driven by ideal input charge packets, as

$$H_{8/8}(z) = \frac{V_{oC}(z)}{q_{iC}(z)} = \frac{k}{(1 - az^{-1}) - e^{j\pi/4}(1 - a)z^{-1}}, \tag{5.18}$$

where k and a are the same as in (5.12) and (5.13). The peak of the transfer function lies at

$$f_{IF} = \frac{f_s}{2\pi} \arctan\left(\frac{(1 - a)\sin(\pi/4)}{a + (1 - a)\cos(\pi/4)}\right). \tag{5.19}$$

The 8/8-phase CS-BPF has a first-order BPF characteristic centered at f_{IF}. In addition to the filtering improvement over its conventional counterpart, this filter is capable of filtering images and out-of-band blockers at third or fifth LO harmonics. It should be noted that this filter still maintains the *full* compatibility with the technology scaling due to its DT passive nature.

5.4.5 8/16-Phase CS-BPF

To further improve the filtering order and characteristics of the 8/8-phase CS-BPF, an IIR LPF (of single or multiple poles) is added during the charge-sharing process in-between every two adjacent inputs. As an example of such a filter, one LPF pole is added between each pair of adjacent input history capacitors, C_H, in Fig. 5.37(f) to give rise to an 8/16-phase CS-BPF. This filter has 8 inputs, 8 outputs, 16 C_H's (8 of them are input C_H's), and 16 nonoverlapped clock phases with a duty cycle of 1/16. The input is interpreted as four *differential* charge packets (q_1, q_2, q_3, and q_4) with multiples of 45° phase shifts provided by the DT mixer. The eight individual *single-ended* input charge packets are accumulated into their respective input C_H's . At the end of each odd-numbered phase $\phi_1, \phi_3, \ldots, \phi_{15}$, the rotating capacitor C_R samples a charge from the active C_H. In the following even-numbered phase of $\phi_2, \phi_4, \ldots, \phi_{16}$, C_R containing the previous packet is charge shared with a newly introduced history capacitor, termed "output C_H," which contains the intermediate (i.e., additionally LPF filtered) version of the "history" charge. Therefore, in each phase, C_R removes a charge proportional to $C_R/(C_H + C_R)$ from each C_H (whether input or output) and then delivers it to the next C_H. The newly introduced output history capacitors add significant extra filtering, thus improving blocker resilience. They also provide convenient pick-up nodes for the dedicated output port that is now physically separate from the input.

5.4.6   General M/N-Phase CS-BPF

In the above case, the 8/16-phase CS-BPF does not operate at the full rate and so all eight outputs can be read out at the maximum sampling rate of $f_s = 1/T_s = f_{LO}$. By defining V_{oC} and q_{iC} the same as (5.12) and (5.12), the filtering transfer function of the filter driven by ideal charge packets, as shown in Fig. 5.37(f), can be proven to be

$$H_{8/16}(z) = \frac{V_{oC}(z)}{q_{iC}(z)} = \frac{k \cdot (1-a)z^{-1}}{(1-az^{-1})^2 - e^{j\pi/4}\left((1-a)z^{-1}\right)^2}, \tag{5.20}$$

where k and a are the same as (5.12) and (5.13), respectively. We find the center frequency of the filter to be

$$f_{IF} = \frac{f_s}{2\pi} \arctan\left(\frac{(1-a)\sin(\pi/8)}{a + (1-a)\cos(\pi/8)}\right). \tag{5.21}$$

To summarize, the blocker-resilient 8/16-phase CS-BPF features a sharp and highly linear transfer function (TF) in order to filter images and out-of-band blockers even at third or fifth harmonics of LO. The out-of-band filtering of blockers is improved significantly compared to [45, 128] by increasing the number of input phases of CS-BPF and adding the LPF pole between each pair of adjacent input history capacitors. The center frequency of the filter is fully controllable by the capacitance ratios and sampling frequency, thus making it insensitive to PVT.

Figure 5.38 presents various configurations of the single-stage full-rate CS-BPF: (a) without the additional LPF poles; (b) with one LPF pole; and (c) with $X = (N/M - 1)$ LPF poles between the adjacent history capacitors. For extending the CS-BPF to a general form, we use the notation of "M/N-phase CS-BPF," where it has M inputs, M outputs, N history capacitors, N nonoverlapped clock phases with a duty cycle of $D = 1/N$, and X LPF poles in the charge-sharing loop. Inputs of the filter are interpreted as differential charge packets, $q_1, q_2, \ldots, q_{M/2}$, that are phase shifted by

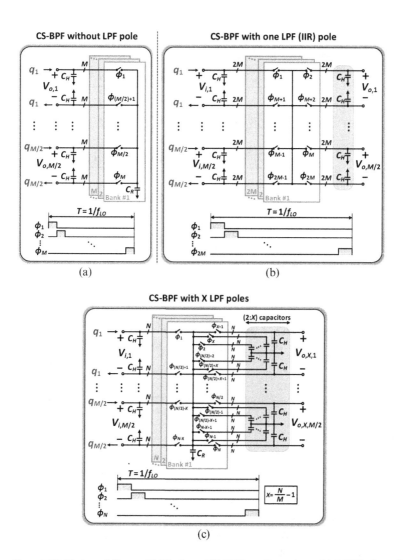

Figure 5.38 Various full-rate M/N-phase CS-BPF configurations. (a) M/M-phase CS-BPF; (b) $M/(2M)$-phase CS-BPF; (c) M/N-phase CS-BPF.

$0, 2\pi/M, 4\pi/M, \ldots, (M-2)\pi/M$ radians and, for the first CS-BPF, provided by the M-phase DT mixer.

To support the full-rate operation, parallelism or interleaving techniques are used to increase the sampling frequency to $f_s = Mf_{LO}$ [128]. As in any sampling system, the frequency components at $f_s \pm f_{IF}$ are folded to the desired frequency at IF. Therefore, larger values of M increase f_s, thus pushing away the closest folding frequencies. Similarly, increasing M improves the CS-BPF tolerance to blockers but at the same time introduces more complexity.

To investigate the transfer function of full-rate M/N-phase CS-BPF, the time-domain output voltage expressions at $t = nT_s$, where $T_s = 1/f_s$, can be derived as

$$V_{i,1}[n] = \frac{C_H V_{i,1}[n-1] + C_R V_{o,X,M/2}[n-1] + 2q_1[n]}{C_H + C_R}, \tag{5.22}$$

$$V_{i,h}[n] = \frac{C_H V_{i,1}[n-1] + C_R V_{o,X,h-1}[n-1] + 2q_h[n]}{C_H + C_R}, \tag{5.23}$$

$$V_{o,2,j}[n] = \frac{C_H V_{o,2,j}[n-1] + C_R V_{i,j}[n-1]}{C_H + C_R}, \tag{5.24}$$

$$V_{o,l,j}[n] = \frac{C_H V_{o,l,j}[n-1] + C_R V_{o,l-1,j}[n-1]}{C_H + C_R}, \tag{5.25}$$

where $i \in [1, M/2], j \in [1, M/2], h \in [2, M/2]$, and $l \in [3, X]$. By performing a conversion from time-domain to z-domain, the general transfer function and center frequency can be derived as

$$H_{M/N}(z) = \frac{\displaystyle\sum_{l=1}^{M/2} (V_{o,X,l}(z)) e^{j(2l-2)\pi/M}}{\displaystyle\sum_{l=1}^{M/2} (q_l(z)) e^{j(2l-2)\pi/M}}$$

$$= \frac{k \cdot \left((1-a)z^{-1}\right)^{\frac{N}{M}-1}}{(1-az^{-1})^{N/M} - e^{j2\pi/M} \left((1-a)z^{-1}\right)^{N/M}}, \tag{5.26}$$

and

$$f_{IF} \approx \frac{f_s}{2\pi} \arctan\left(\frac{(1-a)\sin(2\pi/N)}{a + (1-a)\cos(2\pi/N)}\right), \tag{5.27}$$

where k and a are the same as (5.13) and (5.13), respectively. The simulated and calculated normalized complex transfer functions are plotted in Fig. 5.39 for the conventional (i.e., 4/4-phase), 8/8-, 8/16-, and 16/32-phase CS-BPF with the following conditions: $C_R = 1\,\text{pF}$ and $f_s = 8\,\text{GHz}$, and the same IF frequency ($f_{IF} = 15\,\text{MHz}$). The switch resistance is assumed to be sufficiently small. Most notably, the filtering characteristic of the M/N-phase filter is improved substantially for higher M. Filter's rejection for far-out frequencies depends on its order. Since both 8/16- and 16/32-phase CS-BPFs have a second-order characteristic, they have the same rejection at far-out frequencies. Nevertheless, the close-in rejection of the 16/32-phase filter is higher than that of 8/16-phase. Moreover, the calculated transfer function based on (5.12) agrees well with simulations.

5.4.7   Harmonic Rejection

The differential mixer driven by a square-wave clock is a linear time-variant circuit that downconverts the desired signal together with undesired interferers at higher LO harmonics. In narrowband receivers, those interferers are not of a major concern

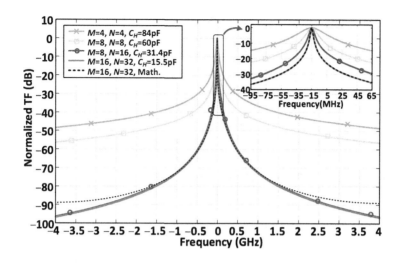

Figure 5.39 Ideal transfer function of M/N-phase CS-BPF.

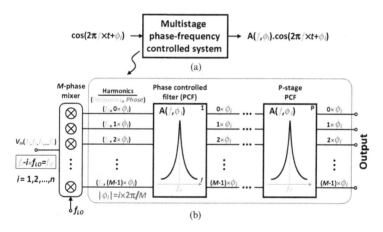

Figure 5.40 Concept of (a) multistage phase-frequency controlled system, (b) multistage phase-controlled filter (PCF).

because of a customary RF band filtering right after the antenna. In *wideband* RF receivers, such RF band select filtering would be very difficult, so it is the LO harmonics instead that need to get rejected. The required level of LO harmonic rejection (HR) is 60–100 dB, which is almost impossible with only one HR stage due to practical amplitude and phase mismatches. A two-stage HR was introduced in [50], but it prevents further HR improvements because of the nonredundant (i.e., quadrature) signal representation. In this section, we introduce a mismatch insensitive HR concept that can be arbitrarily cascaded without any bound on the HR capability.

Figure 5.40(a) starts with a high-level model of a multistage phase-frequency control system. Its key feature is that the harmonic transfer function depends on *both* the input frequency f and phases $\phi_i, i = 0, 1, 2, \ldots$. Multiple phases ϕ_i can be generated with an M-phase mixer, shown in Fig. 5.40(b), which not only down-converts the desired signal at the fundamental but also does the interferers at higher $3\text{rd}, 5\text{th}, \ldots, n\text{th}$ LO harmonics to the same IF frequency with multiple phases of $|\phi_i| = (i - 1) \times 2\pi/M$ where $i = 1, 2, \ldots, M$. Therefore, instead of storing the harmonic information in the frequency domain, as is the case before the mixer $(f_1, f_3, f_5, \ldots, f_n)$, it is now stored as phases in the M mixer output lines, with $M > 4$ to ensure redundancy, where it will be preserved as long as the number of lines is maintained. The multiple phases in M lines can be processed by the phase-controlled filter (PCF), leading to a different transfer function for every harmonic.

5.4.8   CS-BPF Harmonic Rejection Concept

In our implementation, the PCF HR circuitry consists of three stages in total, as shown in Fig. 5.41. It includes two stages of CS-BPFs. Although the first and third or fifth input harmonics are downconverted to the same IF frequency by the octal mixer, the phase difference between two adjacent lines for the first and third or fifth harmonics are $\pi/4$ and $(-3\pi/4)/(5\pi/4)$, respectively. The charge-sharing phases of the signal for the first, third, and fifth harmonics are shown in Fig. 5.41(a). Assuming the even harmonics are removed due to the differential configuration, the phase difference of odd harmonics is sensed by CS-BPF and so the general harmonic TF of the M/N-phase CS-BPF and ϕ_i can be found as

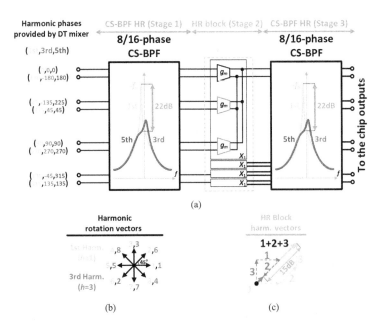

(a)

(b)   (c)

Figure 5.41 (a) Harmonic rejection stages in the superheterodyne receiver, (b) harmonic rotation vectors, and (c) harmonics cancelation summation.

$$H(z, \phi_i) = \frac{1/(C_R + C_H) \cdot \left((1 - a)z^{-1}\right)^{\frac{N}{M} - 1}}{(1 - az^{-1})^{N/M} - e^{j\phi_i} \cdot ((1 - a)z^{-1})^{N/M}}, \tag{5.28}$$

$$\phi_i = (-1)^{\frac{i-1}{2}} \times i \times 2\pi/M, \tag{5.29}$$

respectively, where $i \in [1, 2, \ldots, n]$, and a is equal to (5.12). Figure 5.41(b) shows the corresponding arrangement of phase rotation vectors. The HR for third or fifth harmonics is ~22 dB for each CS-BPF, which can be infinitely improved by cascading CS-BPFs since the octal format fully preserves the harmonic information. HR is further improved by the presented "stage-2" HR block. It consists of four X_1 blocks, each comprising three *identical* g_m-cells adding three adjacent vectors. This results in an amplification of first and partial rejection of the third or fifth harmonic vectors, as shown in Fig. 5.41(c). The two introduced techniques are mismatch insensitive and do not require any calibration, whereas other well-known approaches, such as HR-mixers [135, 11, 136, 50, 19], suffer from such sensitivity so they require

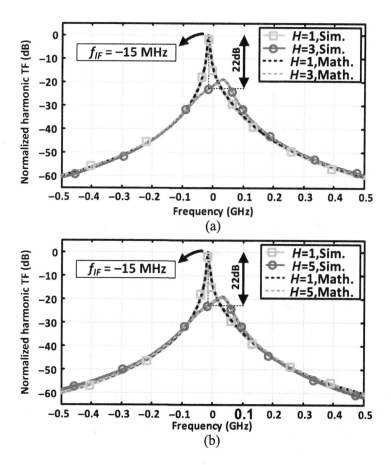

Figure 5.42 Transfer functions of 8/16-phase CS-BPF for different harmonics, both calculated and simulated.

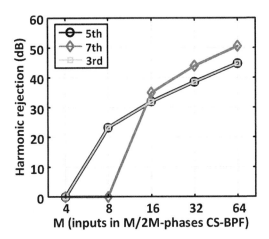

Figure 5.43 Harmonic rejection of 8/16-phase CS-BPF for different harmonics versus M.

extensive calibration. In addition, HR-mixers and switched-capacitor HR [48] cannot be further enhanced because the combined output signals are converted to I/Q (quadrature), thus causing irreversible aliasing of the harmonic phase information.

The simulated normalized transfer functions of the first, third, and fifth harmonics are compared in Fig. 5.42 with calculations based on (5.21). The following conditions apply: $C_R = 1$ pF, $C_H = 31.4$ pF and $f_s = 8$ GHz. The plots verify that the third and fifth harmonics are attenuated by 22 dB. Furthermore, based on (5.21), the third, fifth, and seventh harmonic rejection levels are plotted in Fig. 5.43 versus the number of inputs M for the $M/2M$-phase CS-BPF. The conditions are: $C_R = 1$ pF, $C_H = 31.4$ pF and $f_s = 8$ GHz.

5.4.9   Design and Implementation of the Receiver Chain

We have described so far the evolution of the M/N-phase charge-sharing band-pass filter (CS-BFP) toward its full exploitation as an image reject filter in the fully integrated SAW-less discrete-time superheterodyne receiver. In this section, we describe a detailed design implementation of the receiver, starting with various operational modes of the fully reconfigurable M/N-phase CS-BPF.

5.4.10   4/16-Phase and 8/16-Phase CS-BPFs

The two implemented CS-BPF filters are each programmed as either quadrature (4/16-phase) or octal (8/16-phase). In either mode, the filter is clocked by 16 nonoverlapped signals with $D = 1/16$ and the filter's center frequency is located at IF with no replicas present. The 16 history, C_H, and 16 rotating, C_R, capacitors in the full-rate CS-BPFs shown in Fig. 5.38(b) and (c), are actually 8 differential

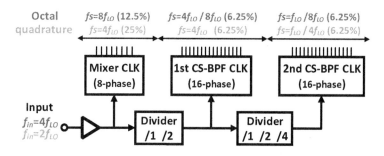

Figure 5.44 Clock generation block diagram.

capacitors each, in order to save the chip area by ×4. Moreover, due to the differential implementation, common-mode voltage and even-order nonlinearity of the prior stages are canceled out. C_H and C_R are digitally tunable with 8-bit binary-weighted codes to support the variable IF of $-10\,\text{MHz}$ up to $-90\,\text{MHz}$ for 2G bands.

5.4.11   Clock Generation Circuitry

Block diagram of the clock generation is shown in Fig. 5.44. An external sinusoidal input is converted to a 50% duty cycle clock after passing through the input buffer. It drives three clock generation circuits. The first circuit provides all the clock phases required for the RF mixer while the remaining two provide all the clock phases for the CS-BPFs. All three circuits are independently programmable to operate in either the octal or quadrature mode. In these modes, the mixer clock generation has a respective output duty cycle of 12.5% and 25%, while the clock for both CS-BPFs is always at $D = 6.25\%$, as shown in Fig. 5.44. To be able to further save the dissipated power, the dividers are used to enable decimation by 1, 2 or 4 for both CS-BPF stages.

Functional block diagram of the clock generation circuitry for the mixer and the two CS-BPF stages is the same. Figure 5.45 shows an example of the mixer LO generation. The CK and $\overline{\text{CK}}$ input clocks with $D = 50\%$ are driving 8 and 4 dynamic latches connected back-to-back in a loop for the octal and quadrature modes, respectively. The latch outputs are followed by digital gates, which produce 12.5% (octal) and 25% (quadrature) duty cycle clocks. The final output is selected between the octal or quadrature outputs by 8 multiplexers. Therefore, in the quadrature mode, half of the mixer switches are off.

The effect of LO phase noise or jitter on the switched-capacitor circuits is discussed in detail in [137, 120, 7]. It can be proven that the CS-BPF and generally passive switch capacitor filters are robust to many nonidealities such as clock jitter, charge injection, nonzero rise/fall times of the clock, and switch resistance [7]. Moreover, there is no need to have a special clocking scheme such as bootstrapped driving and dummy switches [7]. Also, it has been shown that integration sampler, IIR, and FIR filters are exceptionally robust to the clock jitter [7]. The results can be further generalized for the CS-BPF.

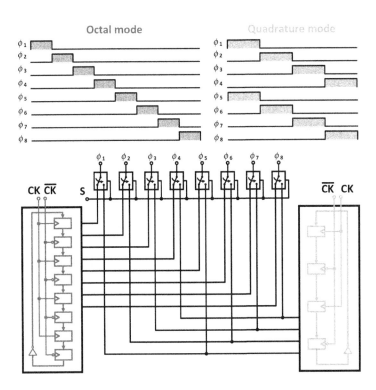

Figure 5.45 Functional block diagram of the mixer clock generation for both octal and quadrature modes.

5.4.12   Low-Noise Transconductance Amplifier (LNTA)

Figure 5.46(a) shows a fully differential schematic of the presented LNTA, which simultaneously features low NF and high IIP3 (only single-ended signal waveforms are shown). The noise-canceling common-gate transistors (M_{n1}/M_{n2}) provide the RX input matching. The noise-canceling operation is as follows: the input signal gets amplified by transistors M_{n1}/M_{n3} and M_{p1} in a differential feed-forward manner, whereas the thermal noise of M_{n1} channel experiences subtraction at the output nodes because of the out-of-phase correlated noise voltages at V_x and V_{outn}. The third-order nonlinearity of M_{n1} and M_{n3} can be simultaneously canceled at the differential output because M_{n1} and M_{n3} operate in weak and saturation regions, respectively, resulting in out-of-phase g_{m3} (third-order transconductance) to each other. Therefore, partial cancelation of the IM3 component happens at the differential output. The cancelation happens at the desired frequency because at other frequencies an additional IM3 is generated due to the second-order nonlinearity of M_{n3}. Simulated (with extracted parasitics) NF and gain of LNTA with a resistive load are shown in Fig. 5.46(b) across 0.1–4 GHz.

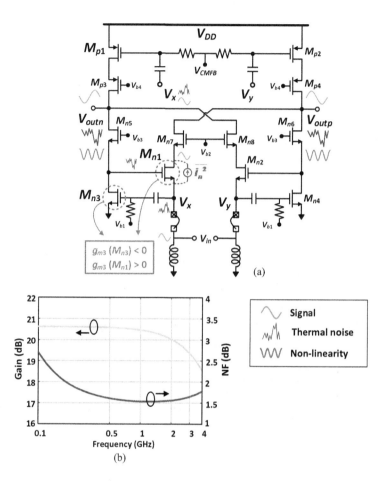

Figure 5.46 (a) LNTA schematic and (b) it post-layout simulated noise figure and gain.

5.4.13   IF Stage Transconductance Amplifier

Figure 5.47 shows a schematic of the pseudo-differential inverter-based IF transconductance amplifier with a common-mode (CM) rejection load. The g_m-cell operates at 0.9 V supply and a pair of complementary thick-oxide PMOS/NMOS transistors is utilized to increase the transconductance linearity to $>+11$dBm (simulated) for all corner cases within a temperature range of $-30°C$ to $100°C$ [61]. The common-mode feedback circuitry provides a proper bias of $V_{DD}/2$ to the outputs.

To suppress any possible CM oscillation in the RX chain, the CM gain of the g_m-cell is drastically reduced by placing a CM load at its output. It features different impedances for the CM and differential-mode (DM) signals. The impedance for DM signals is very high; it is proportional to the small-signal drain resistance of the CM load transistors M_n and M_p, while the impedance for CM signals is very low, equal to $1/((g_{mn} + g_{mp})A)$, where g_{mn} and g_{mp} are the small-signal transconductance of M_n and M_p.

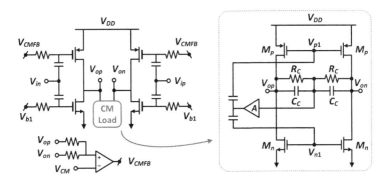

Figure 5.47 The IF g_m-cell schematic with common-mode rejection load.

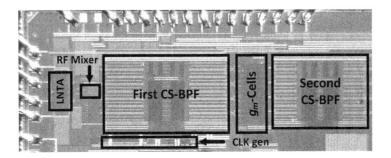

Figure 5.48 Chip micrograph of the introduced discrete-time superheterodyne receiver.

5.4.14   Measurement Results

Figure 5.48 shows the chip micrograph of the presented superheterodyne RX for 4G cellular mobiles realized in TSMC 28 nm CMOS [15]. The active area is $0.52\,\text{mm}^2$, which is mostly occupied by C_H and C_R capacitors of the two CS-BPFs. Both the receiver and clock inputs are differential and so wideband "hybrids" are used to interface with $50\,\Omega$ single-ended instrumentation. All the measurements are performed at high RX gain without any calibrations, even those concerning the linearity. The chip is wire-bonded to a printed circuit board (PCB) providing dc and RF input connectivity ports, while high-IF output signals are measured with a high performance oscilloscope, as shown in Fig. 5.49 and the characteristics of PCB lines and cables are de-embedded from the measurement results. The transfer function measurement setup of the RF CS-BPF is shown in Fig. 5.49. After providing the proper power supply voltages, the serial peripheral interface (SPI) controls internal registers. The quadrature (I/Q) IF outputs are connected to high performance "DSO-X 3052A" digital oscilloscope.

MATLAB scripts are developed to make the chip testing automatic or semiautomatic. The graphical user interface (GUI), shown in Fig. 5.50, are designed to facilitate the testing process and visualize the results of close-in transfer function, wideband transfer function, IIP2, IIP3, and CP linearity measurements. The LO frequency and

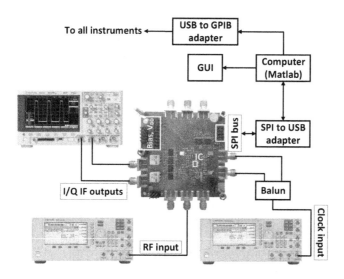

Figure 5.49 Chip micrograph of the introduced discrete-time superheterodyne receiver.

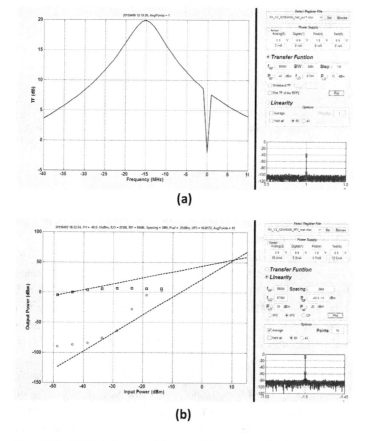

(a)

(b)

Figure 5.50 GUI interface for chip testing for: (a) transfer function and (b) linearity measurements.

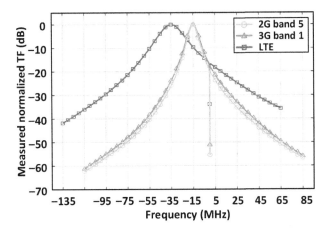

Figure 5.51 Measured RX transfer function for different bands.

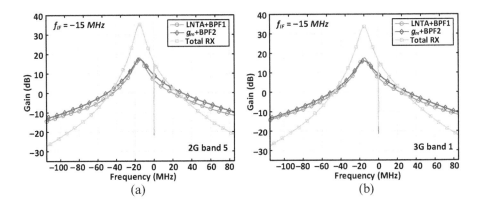

Figure 5.52 Measured RX gain versus output frequency.

RF input frequency are applied to the chip through GPIB connection, and FFT of the output I/Q IF signals has been taken in the MATLAB script.

The measured normalized transfer functions are shown in Fig. 5.51 for "2G band-5," "3G band-1," and "LTE" with 0.85, 2.1, and 2.5 GHz RF input frequencies. The RX bandwidth is 6.5 MHz for 2G/3G and 20 MHz for LTE, while IF frequency is -15 MHz and -35 MHz for 0.85–2.1 GHz and 2.5 GHz carriers, respectively. Moreover, the absolute value of IF in this receiver can be variable in the face of a large blocker, within the range of 10–90 MHz, 25–220 MHz, and 29–262 MHz for 2G, 3G, and LTE bands, respectively.

Figure 5.52 shows the RX gain at 0.85 GHz and 2.1 GHz carriers for I channel only. By recombining the I/Q channels, an extra 6 dB gain can be obtained. The overall passband gain of LNTA and first CS-BPF in 2G band-5 and 3G band-1 is around 18 dB and 17.5 dB, respectively. The gain of IF g_m-cell and second CS-BPF is measured by subtracting the total RX gain from the gain provided by LNTA and first

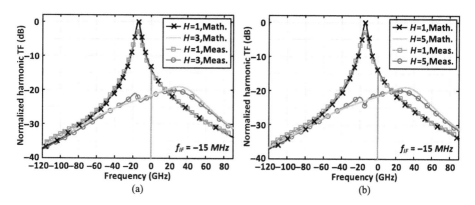

Figure 5.53 Comparison of the normalized measured first, third, and fifth harmonic TF with calculation for 2G band. All transfer functions are normalized to maximum gain of first harmonic extracted from the calculation.

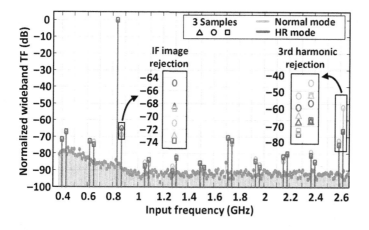

Figure 5.54 Measured wideband transfer function of the complete RX.

CS-BPF. That peak gain value is 17 dB and 16.5 dB for 2G and 3G, respectively. The total RX gain is between 29 and 35 dB for 0.85–2.5 GHz carriers. Although the first and second CS-BPFs are identical, the former shows a sharper filtering characteristic due to a larger output resistance of LNTA versus that of IF g_m-cell.

The comparisons of the measured transfer functions of LNTA and first CS-BPF with calculations per (5.28) are shown in Fig. 5.53(a) and (b), respectively, for third and fifth harmonics. The difference between the measured and calculated first harmonic at IF is due to the effect of LNTA output impedance. The 19 dB rejection of third and fifth harmonics per each CS-BPF stage is measured at IF.

The measured wideband transfer functions in the normal and HR modes for three ICs is shown in Fig. 5.54. All the images are rejected by more than 65 dB, including the IF image, in all three measured ICs without any calibration. The worst-case HR of 58 dB is achieved when the HR block is enabled: 38 dB from the two-stage

Table 5.4. Noise figure contribution of each building block in the RX chain

	LNTA	RF mixer	CS-BPF1	IF g_m-cell	CS-BPF2
2G band-5 (%)	89.27	0	4.97	4.98	0.78
3G band-1 (%)	84.89	0	6.86	6.81	1.44

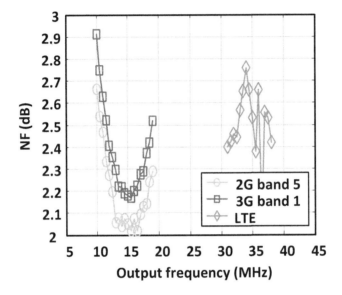

Figure 5.55 Measured noise figure for 2G, 3G, and LTE bands.

CS-BPFs, 17 dB from the HR block, and the rest is provided by the LNTA's limited bandwidth. The highlighted images are multiples of the smallest LO frequency in the clock generation circuitry with an offset of $\pm f_{IF}$.

Figure 5.55 plots the measured receiver NF of 2.1–2.6 dB with an LO frequency of 865 MHz, 2115 MHz, and 2535 MHz for 2G, 3G, and LTE, respectively. The minimum noise figure in each standard happens at the center frequency of CS-BPFs, which coincides with the IF location. Furthermore, the NF contribution of each building block is summarized in Table 5.4 for 2G band-5 and 3G band-1.

The simulated (post-layout extracted) out-of-band IIP3 of CS-BPF is more than +30 dBm. Furthermore, because of its strong blocker filtering, out-of-band IIP3 is mainly determined by the linearity of LNTA. Figure 5.56 shows the measured out-of-band IIP3 of the RX versus offset frequency for 2G and 3G. It should be mentioned that the linearity was measured at the maximum gain (i.e., the lowest noise figure) and without any calibration. The variation of out-of-band IIP3 over the offset frequency is due to the linearity dependency of LNTA on the offset frequency. The peak IIP3 of +14 dBm is achieved for the offset frequencies specified by the 2G/3G standards at duplex (f_{TX}) and half-duplex (($f_{TX} + f_{RX}$)/2) frequencies.

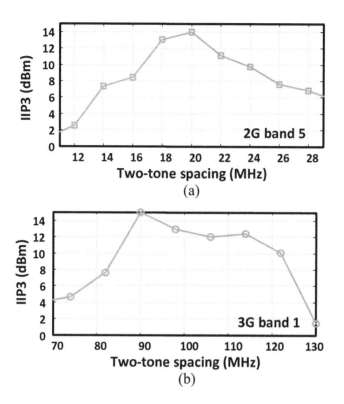

Figure 5.56 Measured IIP3 for (a) 2G and (b) 3G bands versus frequency offset.

For the IIP2 measurements, there are several IIP2 test cases that the two most important ones are:

1. closely spaced tones or a modulated single tone IIP2 test case (limitation in mixer IIP2) and
2. far away two-tone cases (limitation in LNA)

The test case that prevents us from removing the SAW filter is the first one since if there were no SAW filter in the RX chain, the IIP2 of more than +90dBm would be needed. The second test case is the one that should be also investigated in wideband RXs. However, it is not actually that *stringent* compared to the first test case. Let us calculate the needed IIP2 for the second test case. Assuming the blocker level of −32.5 dBm applied to the RX, if we need a sensitivity of −99 dBm and SNR of 9 dB to maintain the signal purity. The IM2 component should be below −108 dBm. Therefore, the needed IIP2 for the second test case is 43 dBm. To clarify the situation, we have performed an IIP2 measurement for both test cases.

For the first test case, since the RX architecture is superheterodyne with an IF frequency of −15 MHz to −35 MHz, the applied two-tone or one modulated tone with 7.5 MHz bandwidth will be downconverted to around dc, thus *completely* filtered out.

For the second test case, the two tones are far away from each other, and the generated IM2 is actually in-band. In this RX, the two tones should be located at

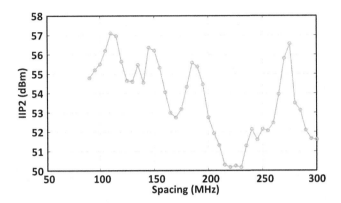

Figure 5.57 Measured IIP2 for far away two-tone case at $f_{RF} + spacing$ and $2f_{RF} + spacing$ while f_{RF} is 860 MHz versus spacing.

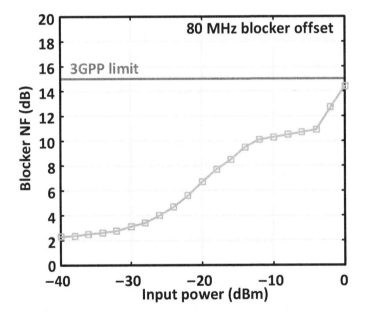

Figure 5.58 Measured blocker noise figure at 80 MHz offset frequency.

$f_{RF} + spacing$ and $2f_{RF} + spacing$ while f_{RF} is 860 MHz in 2G band-5. As shown in Fig. 5.57, the IIP2 more than +50 dBm is achieved in a wide frequency spacing when the LNTA is set to mid gain because in that case there is no need for LNTA to be at the high gain setting.

The RX blocker tolerance is demonstrated by means of the blocker NF tests. Figure 5.58 shows the NF as a function of the 80 MHz blocker power. NF remains below the 15 dB limit for the ≤ 0 dBm blocker. Furthermore, an external BPF is used to reduce the impact of the LO generation phase noise on the input RF band.

The measured power consumption of the RX chip versus input frequency is shown in Fig. 5.59. The overall RX power consumption varies from 22 to 40 mW dependent

Table 5.5. Performance summary and comparison with state-of-the-art

	This work	[8] JSSC '11 Broadcom	[45] ISSCC '14 TUDelft	[19] JSSC '14 IMEC	[138] ISSCC '14 Broadcom	[18] ISSCC '14 MediaTek	[27] JSSC '13 UPavia
Architecture		Superheterodyne				Zero-IF	
CMOS Tech. (nm)	28	65	65	28	28	40	40
SAW-less	Yes	No	No	Yes	Yes	Yes	Yes
RF input	Differential	Single-ended	Differential	Differential	Differential	Differential	Differential/Single-ended
RF frequency (GHz)	0.5–2.5	1.8–2.2	1.8–2.5	0.4–6	0.1–3.3	0.85–2.4	1.8–2.4
NF (dB)[e]	2.1/2.2/2.6	2.8	3.2–4.5	(1.8–3)(4.6^d)	1.7	1.7–2.4	3.8/1.9
Supply (V)	0.9	1.2/2.5	1.2/2.0	0.9	1.0	1.5	1.2/1.8
HR (dB)	>58[a]	N/A	N/A	50(70^b)	60/60	N/A	54/65
OB-IIP3 (dBm)	14	−6.3	–	8/5	11.5	0.4	18/16
IB-IIP3 (dBm)	−10	−8.4	−7	+4[g]	–	1.9	–
OB-IIP2 (dBm)	Infinite[f]	N/A	85[c]	55(88^b)	55	55	64
OB-Blocker NF (dB)	14	N/A	N/A	14(10^d)	5	11	7.9
Image Rej. (dB)	>65[a]	35	37	N/A	N/A	N/A	N/A
S11 (dB)	<−10	<−10	<−10	<−10	<−10	<−10	<−10
Power (mW)	22/35/40	39	55–65	40/35	36–62	39–46.5	32/32
IF (MHz)	2G: 10–90 3G: 25–220 LTE: 29–262	110	27–200	0	0	0	0
Area (mm²)	0.52	0.76	1.1	0.6	5.2	0.57	0.84/0.74

[a] Worst-case without calibration and measured 3 IC samples, [b] with calibration, [c] due to an IF mixer (second mixer), [d] with optimized setting, [e] 1dB typical input balun loss should be included in TDD mode for RXs with differential inputs, [f] for closely spaced or modulated interferers, [g] Reduced gain setting

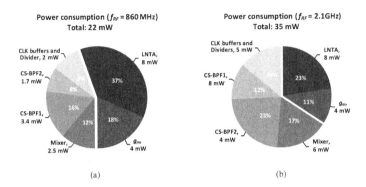

Figure 5.59 Measured RX power consumption for (a) 2G band-5 ($f_{RF} = 860$ MHz); and (b) 3G band-1 ($f_{RF} = 2.1$ GHz) carriers.

on the input RF band and related clock frequency. The main contributor to the overall RX power is the analog part for "2G band-5." As the clock frequency increases for "3G band-1," the main contributor is the power consumed by the DT part including RF mixer, CS-BPF1, CS-BPF2, and clock buffers and dividers.

Table 5.5 compares the introduced DT RX with state-of-the-art RXs. While being the best-in-class in meeting the key performance parameters without any calibration, its power consumption and area are generally the lowest, and it does not suffer from any issues related to dc offsets, flicker noise, or IM2 products since its IIP2 is infinite.

5.5   Conclusion

Four representative examples of the discrete-time RF cellular receivers, including zero-IF and superheterodyne, are presented in this chapter. The first example uses the RF direct sampling technique to achieve great selectivity right at the mixer level. The dynamic range requirements of the following ADC can thus be significantly relaxed. This technique of multi-tap direct sampling mixer (MTDSM) was validated in commercial Bluetooth and GSM radios.

In the second architecture, the concept of impedance combination was utilized to realize the complex high-Q RF BPF that rejects the image folding, which has prevented the widespread adoption of high-IF RX architectures in the past.

The third example shows the complete chain of discrete-time (DT) superheterodyne receiver with high reconfigurability. The full monolithic integration is made possible by the novel DT BPF.

Finally, the superheterodyne receiver that can be finally claimed as SAW-less is introduced. The consequence of reduced filtering at the antenna interface network forces much better linearity and filtering requirements in the on-chip RF front-end. The BPF with several extra LPF poles in charge-sharing rotation path is tolerant of blockers, making it a suitable candidate not only for SAW-less superheterodyne receivers, but also for low-IF RXs. The architecture was realized in 28 nm CMOS and is amenable to further scaling.

References

[1] M. Tohidian, I. Madadi, and R. B. Staszewski. A fully integrated discrete-time super-heterodyne receiver with +90 dBm uncalibrated IIP2. In *Proceedings of the IEEE International Solid-State Circuits Conference*, 2015.

[2] C. P. Yue and S. S. Wong. Scalability of RF CMOS. In *Proceedings of the 2005 IEEE Radio Frequency Integrated Circuits (RFIC) Symposium – Digest of Papers*, pages 53–56, 2005.

[3] C. H. Diaz, D. D. Tang, and J. Y. C. Sun. CMOS technology for MS/RF SoC. *IEEE Transactions on Electron Devices*, 50(3):557–566, 2003.

[4] M. Tohidian, I. Madadi, and R. Bogdan Staszewski. A 2mW 800MS/s 7th-order discrete-time IIR filter with 400kHz-to-30MHz BW and 100dB stop-band rejection in 65nm CMOS. In *Proceedings of the IEEE International Solid-State Circuits Conference – Digest of Technical Papers*, volume 56, pages 174–175, 2013.

[5] M. Tohidian, I. Madadi, and R. B. Staszewski. Analysis and design of a high-order discrete-time passive IIR low-pass filter. *IEEE Journal of Solid-State Circuits*, 49(11): 2575–2587, 2014.

[6] I. Madadi, M. Tohidian, and R. B. Staszewski. A 65nm CMOS high-IF superheterodyne receiver with a High-Q complex BPF. In *Proceedings of the 2013 IEEE Radio Frequency Integrated Circuits (RFIC) Symposium*, pages 323–326, 2013.

[7] A. Mirzaei, S. Chehrazi, R. Bagheri, and A. A. Abidi. Analysis of first-order anti-aliasing integration sampler. *IEEE Transactions on Circuits and Systems I*, 55(10):2994–3005, 2008.

[8] A. Mirzaei, H. Darabi, and D. Murphy. A low-power process-scalable super-heterodyne receiver with integrated high-Q filters. *IEEE Journal of Solid-State Circuits*, 46(12):2920–2932, 2011.

[9] R. B. Staszewski, K. Muhammad, D. Leipold et al. All-digital TX frequency synthesizer and discrete-time receiver for Bluetooth radio in 130-nm CMOS. *IEEE Journal of Solid-State Circuits*, 39(12):2278–2291, 2004.

[10] K. Muhammad, Y. Ho, T. L. Mayhugh et al. The first fully integrated quad-band GSM/GPRS receiver in a 90-nm digital CMOS process. *IEEE Journal of Solid-State Circuits*, 41(8):1772–1783, 2006.

[11] R. Bagheri, A. Mirzaei, S. Chehrazi et al. An 800-MHz-6-GHz software-defined wireless receiver in 90-nm CMOS. *IEEE Journal of Solid-State Circuits*, 41(12):2860–2875, 2006.

[12] A. Geis, J. Ryckaert, L. Bos et al. A 0.5 mm^2 power-scalable 0.5–3.8-GHz CMOS DT-SDR receiver with second-order RF band-pass sampler. *IEEE Journal of Solid-State Circuits*, 45(11):2375–2387, 2010.

[13] R. Chen and H. Hashemi. A 0.5-to-3 GHz software-defined radio receiver using discrete-time RF signal processing. *IEEE Journal of Solid-State Circuits*, 49(5):1097–1111, 2014.

[14] I. Madadi, M.Tohidian, and R. B. Staszewski. In Press A high IIP2 SAW-less super-heterodyne receiver with multi-stage harmonic rejection. *IEEE Journal of Solid-State Circuits*, 51:332–347, 2016.

[15] I. Madadi, M.Tohidian, and R. B. Staszewski. A TDD/FDD SAW-less superheterodyne receiver with blocker-resilient band-pass filter and multi-stage HR in 28nm CMOS. In *2015 IEEE Symposium on VLSI Circuits – Digest of Technical Papers, pages 1–2*, 2015.

[16] T. H. Lee. *The Design of CMOS Radio Frequency Integrated Circuits*, 2nd ed., Cambridge University Press, 2004.

[17] I. Fabiano, M. Sosio, A. Liscidini, et al. SAW-Less analog front-end receivers for TDD and FDD. *IEEE Journal of Solid-State Circuits*, 48(12):3067–3079, 2013.

[18] M. D. Tsai, C. F. Liao, C. Y. Wang et al. A multi-band inductor-less SAW-less 2G/3G-TD-SCDMA cellular receiver in 40nm CMOS. In *Proceedings of the IEEE International Solid-State Circuits Conference – Digest of Technical Papers*, volume 57, pages 354–355, 2014.

[19] B. van Liempd, J. Borremans, E. Martens et al. A 0.9 V 0.4–6 GHz harmonic recombination SDR receiver in 28 nm CMOS with HR3/HR5 and IIP2 calibration. *IEEE Journal of Solid-State Circuits*, 49(8):1815–1826, 2014.

[20] D. Murphy, A. Mirzaei, H. Darabi et al. An LTV analysis of the frequency translational noise-cancelling receiver. *IEEE Transactions on Circuits and Systems I*, 61(1):266–279, 2014.

[21] D. Murphy, H. Darabi, A. Abidi et al. A blocker-tolerant, noise-cancelling receiver suitable for wideband wireless applications. *IEEE Journal of Solid-State Circuits*, 47(12):2943–2963, 2012.

[22] D. Kaczman, M. Shah, M. Alam et al. A single-chip 10-band WCDMA/HSDPA 4-band GSM/EDGE SAW-less CMOS receiver with DigRF 3G interface and +90 dBm IIP2. *IEEE Journal of Solid-State Circuits*, 44(3):718–739, 2009.

[23] B. Razavi. *Fundamentals of Microelectronics*, 2nd ed. Wiley, 2013

[24] B. van Liempd, J. Borremans, S. Cha, E. Martens, H. Suys, and J. Craninckx. IIP2 and HR calibration for an 8-phase harmonic recombination receiver in 28nm. In *Proceedings of IEEE Custom Integrated Circuits Conference*, pages 2–5, 2013.

[25] B. Debaillie, P. van Wesemael, G. Vandersteen et al. Calibration of direct-conversion transceivers. *IEEE Journal of Selected Topics in Signal Processing*, 3(3):488–498, 2009.

[26] Digital cellular telecommunications system (Phase 2+); Radio transmission and reception. ETSI TS 100 910 V8.6.0 (2000-09), 2013.

[27] I. Fabiano, M. Sosio, A. Liscidini, and R. Castello. SAW-less analog front-end receivers for TDD and FDD. *IEEE Journal of Solid-State Circuits*, 48(12):3067–3079, 2013.

[28] S. Chehrazi, A. Mirzaei, and A. A. Abidi. Second-order intermodulation in current-commutating passive FET mixers. *IEEE Transactions on Circuits and Systems I*, 56(12):2556–2568, 2009.

[29] E. E. Bautista, B. Bastani, and J. Heck. High IIP2 downconversion mixer using dynamic matching. *IEEE Journal of Solid-State Circuits*, 35(12):1934–1941, 2000.

[30] M. Brandolini, P. Rossi, D. Sanzogni, and F. Svelto. A +78 dBm IIP2 CMOS direct downconversion mixer for fully integrated UMTS receivers. *IEEE Journal of Solid-State Circuits*, 41(3):552–559, 2006.

[31] H. Darabi, H. J. Kim Hea Joung Kim, J. Chiu, B. Ibrahim, and L. Serrano. An IP2 improvement technique for zero-IF down-converters. In *Proceedings of the 2006 IEEE International Solid-State Circuits Conference – Digest of Technical Papers*, volume 38, pages 171–174, 2006.

[32] I. Elahi and K. Muhammad. IIP2 calibration by injecting DC offset at the mixer in a wireless receiver. *IEEE Transactions on Circuits and Systems II*, 54(12):1135–1139, 2007.

[33] K. Dufrene and R. Weigel. A novel IP2 calibration method for low-voltage downconversion mixers. In *Proceedings of the 2006 IEEE Radio Frequency Integrated Circuits (RFIC) Symposium*, 2006.

[34] Q. Huang, J. Rogin, X. Chen et al. A tri-band SAW-less WCDMA/HSPA RF CMOS transceiver with on-chip DC-DC converter connectable to battery. In *Proceedings of the IEEE International Solid-State Circuits Conference – Digest of Technical Papers*, volume 53, pages 60–61, 2010.

[35] Y. Feng, G. Takemura, S. Kawaguchi, N. Itoh, and P. Kinget. A low-power low-noise direct-conversion front-end with digitally assisted IIP2 background self calibration. In *Proceedings of the IEEE International Solid-State Circuits Conference – Digest of Technical Papers*, volume 53, pages 70–71, 2010.

[36] K. Dufrêne, Z. Boos, and R. Weigel. A 0.13μm 1.5V CMOS I/Q downconverter with digital adaptive IIP2 calibration. In *Proceedings of the IEEE International Solid-State Circuits Conference – Digest of Technical Papers*, volume 2, pages 80–82, 2007.

[37] K. Dufrêne, Z. Boos, and R. Weigel. Digital adaptive IIP2 calibration scheme for CMOS downconversion mixers. *IEEE Journal of Solid-State Circuits*, 43(11):2434–2445, 2008.

[38] Y. Feng, G. Takemura, S. Kawaguchi, N. Itoh, and P. R. Kinget. Digitally assisted IIP2 calibration for CMOS direct-conversion receivers. *IEEE Journal of Solid-State Circuits*, 46(10):2253–2267, 2011.

[39] M. Chen, Y. Wu, and M. F. Chang. Active 2nd-order intermodulation calibration for direct-conversion receivers. *In Proceedings of the IEEE International Solid-State Circuits Conference – Digest of Technical Papers*, volume 38(6), pages 171–174, 2006.

[40] L. Longo, R. Halim, B.-R. Horng et al. A cellular analog front end with a 98 dB IF receiver. In *Proceedings of the 1994 IEEE International Solid-State Circuits Conference*, pages 226–227, 1994.

[41] A. Hairapetian. An 81-MHz IF receiver in CMOS. *IEEE Journal of Solid-State Circuits*, 31(12):1981–1986, 1996.

[42] B. Staszewski, K.Muhammad, and D. Leipold. A discrete-time Bluetooth receiver in a 0.13μm digital CMOS process. In *Proceedings of the IEEE International Solid-State Circuits Conference*, pages 268–269, Feb 2004.

[43] K. Muhammad and R. B. Staszewski. Direct RF sampling mixer with recursive filtering in charge domain. *Proceedings of the 2004 IEEE International Symposium on Circuits and Systems*, pages 577–580, May 2004.

[44] R. B. Staszewski and K. Muhammad. Joint common mode voltage and differential offset voltage control scheme in a low-IF receiver. *Proceedings of the 2004 IEEE Radio Frequency Integrated Circuits (RFIC) Symposium*, pages 405–408, June 2004.

[45] M. Tohidian, I. Madadi, and R. B. Staszewski. A fully integrated highly reconfigurable discrete-time superheterodyne receiver. In *Proceedings of the 2014 IEEE International Solid-State Circuits Conference*, pages 72–74, 2014.

[46] S. Karvonen, T. A. D. Riley, and J. Kostamovaara. A CMOS quadrature charge-domain sampling circuit with 66-dB SFDR up to 100 MHz. *IEEE Transactions on Circuits and Systems I*, 52(2):292–304, 2005.

[47] S. Karvonen. Charge-domain sampling of high-frequency signals with embedded filtering. PhD thesis, University of Oulu, 2006.

[48] Z. Ru, E. A. M. Klumperink, and B. Nauta. Discrete-time mixing receiver architecture for RF-sampling software-defined radio. *IEEE Journal of Solid-State Circuits*, 45(9):1732–1745, 2010.

[49] F. Bruccoleri, E. A. M. Klumperink, and B. Nauta. Wide-band CMOS low-noise amplifier exploiting thermal noise canceling. *IEEE Journal of Solid-State Circuits*, 39(2):275–282, February 2004.

[50] Z. Ru, N. A. Moseley, E. A. M. Klumperink, and B. Nauta. Digitally enhanced software-defined radio receiver robust to out-of-band interference. *IEEE Journal of Solid-State Circuits*, 44(12):3359–3375, 2009.

[51] X. Wang, J. Sturm, N. Yan, X. Tan, and H. Min. 0.6–3-GHz wideband receiver RF front-end with a feedforward noise and distortion cancellation resistive-feedback LNA. *IEEE Transactions on Microwave Theory and Techniques*, 60(2):387–392, 2012.

[52] A. Bevilacqua and A. M. Niknejad. An ultrawideband CMOS low-noise amplifier for 3.1–10.6-GHz wireless receivers. *IEEE Journal of Solid-State Circuits*, 39(12):2259–2268, 2004.

[53] Z. Li, L. Chen, Z. Wang, et al. Low-noise and high-gain wideband LNA with g_m-boosting technique. *Electronics Letters*, 49(18):1126–1128, 2013.

[54] H. Zhang, X. Fan, and E. S. Sinencio. A low-power, linearized, ultra-wideband LNA design technique. *IEEE Journal of Solid-State Circuits*, 44(2):320–330, 2009.

[55] K. C. He, M. T. Li, C. M. Li, and J. H. Tarng. Parallel-RC feedback low-noise amplifier for UWB applications. *IEEE Transactions on Circuits and Systems II: Express Briefs*, 57(8):582–586, 2010.

[56] S. B. T. Wang, A. M. Niknejad, and R. W. Brodersen. Design of a sub-mW 960-MHz UWB CMOS LNA. *IEEE Journal of Solid-State Circuits*, 41(11):2449–2456, 2006.

[57] M. Moezzi and M. Sharif Bakhtiar. Wideband LNA using active inductor with multiple feed-forward noise reduction paths. *IEEE Transactions on Microwave Theory and Techniques*, 60(4):1069–1078, 2012.

[58] F. Zhang and P. R. Kinget. Low-power programmable gain CMOS distributed LNA. *IEEE Journal of Solid-State Circuits*, 41(6):1333–1343, 2006.

[59] D. Murphy, A. Hafez, A. Mirzaei, M. Mikhhemar, H. Darabi, M. Chang, and A. Abidi. A blocker-tolerant wideband noise-canceling receiver with a 2-dB noise figure. In *Proceedings of the IEEE International Solid-State Circuits Conference (ISSCC)*, pages 74–76, 2012.

[60] S. C. Blaakmeer, E. A. M. Klumperink, D. M. W. Leenaerts, and B. Nauta. Wide-band balun-LNA with simultaneous output balancing, noise-canceling and distortion-canceling. *IEEE Journal of Solid-State Circuits*, 43(6):1341–1350, 2008.

[61] H. Zhang and E. Sanchez-Sinencio. Linearization techniques for CMOS low noise amplifiers: A tutorial. *IEEE Transactions on Circuits and Systems I: Regular Papers*, 58(1):22–36, 2011.

[62] P. Simitsakis, Y. Papananos, and E. S. Kytonaki. Design of a low voltage-low power 3.1–10.6 GHz UWB RF front-end in a CMOS 65 nm technology. *IEEE Transactions on Circuits and Systems II: Express Briefs*, 57(11):833–837, 2011.

[63] A. Bozorg and R. B. Staszewski. A 0.02–4.5-GHz LN(T)A in 28-nm CMOS for 5G exploiting noise reduction and current reuse. *IEEE Journal of Solid-State Circuits*, 55:404–415 2021.

[64] Y. J. Lin, S. S. H. Hsu, J. D. Jin, and C. Y. Chan. A 3.1–10.6 GHz ultra-wideband CMOS low noise amplifier with current-reused technique. *IEEE Microwave and Wireless Components Letters*, 17(3):232–234, 2007.

[65] R. M. Weng, C. Y. Liu, and P. C. Lin. A low-power full-band low-noise amplifier for ultra-wideband receivers. *IEEE Transactions on Microwave Theory and Techniques*, 58(8):2077–2083, 2010.

[66] A. Ansari and M. Yavari. A very wideband low noise amplifier for cognitive radios. In *2011 18th IEEE International Conference on Electronics, Circuits, and Systems*, pages 623–626, 2012.

[67] I. Madadi, M. Tohidian, K. Cornelissens, P. Vandenameele, and R. B. Staszewski. A high IIP2 SAW-less superheterodyne receiver with multistage harmonic rejection. *IEEE Journal of Solid-State Circuits*, 51(2):332–347, 2016.

[68] Y.-C. Ho, R. B. Staszewski, K. Muhammad, C.-M. Hung, D. Leipold, and K. Maggio. Charge-domain signal processing of direct RF sampling mixer with discrete-time filters in Bluetooth and GSM receivers. *EURASIP Journal on Wireless Communications and Networking*, 2006(1):1–14, 2006.

[69] C. F. Liao and S. I. Liu. A broadband noise-canceling CMOS LNA for 3.1–10.6-GHz UWB receivers. *IEEE Journal of Solid-State Circuits*, 42(2):329–339, 2007.

[70] B. Toole, C. Plett, and M. Cloutier. RF circuit implications of moderate inversion enhanced linear region in MOSFETs. *IEEE Transactions on Circuits and Systems I: Regular Papers*, 51(2):319–328, 2004.

[71] B. Razavi. *RF Microelectronics*, 2nd edition. Prentice Hall Press, 2011.

[72] S. Kim and K. Kwon. Broadband balun-LNA employing local feedback g_m-boosting technique and balanced loads for low-power low-voltage applications. *IEEE Transactions on Circuits and Systems I: Regular Papers*, pages 1–10, 2020.

[73] H. Yu, Y. Chen, C. C. Boon, P.-I. Mak, and R. P. Martins. A 0.096-mm^2 1–20-GHz triple-path noise-canceling common-gate common-source LNA with dual complementary pMOS–nMOS configuration. *IEEE Transactions on Microwave Theory and Techniques*, 68(1):144–159, 2020.

[74] S. Kim and K. Kwon. A 50 MHz – 1 GHz 2.3-dB NF noise-cancelling balun-LNA employing a modified current-bleeding technique and balanced loads. *IEEE Transactions on Circuits and Systems I*, 66(2):546–554, 2019.

[75] S. S. Regulagadda, B. D. Sahoo, A. Dutta, K. Y. Varma, and V. S. Rao. A packaged noise-canceling high-gain wideband low noise amplifier. *IEEE Transactions on Circuits and Systems II: Express Briefs*, 66(1):11–15, 2019.

[76] J. Jang, H. Kim, G. Lee, and T. W. Kim. Two-stage compact wideband flat gain low-noise amplifier using high-frequency feedforward active inductor. *IEEE Transactions on Microwave Theory and Techniques*, pages 1–9, 2019.

[77] H. Yu, Y. Chen, C. C. Boon, C. Li, P.-I. Mak, and R. P. Martins. A 0.044-mm2 0.5-to-7-GHz resistor-plus-source-follower-feedback noise-cancelling LNA achieving a flat NF of 3.3–0.45 dB. *IEEE Transactions on Circuits and Systems II: Express Briefs*, 66(1):71–75, 2019.

[78] Z. Pan, C. Qin, Z. Ye, Y. Wang, and Z. Yu. Wideband inductorless low-power LNAs with g_m enhancement and noise-cancellation. *IEEE Transactions on Circuits and Systems I*, 65(1):26–38, 2018.

[79] A. R. Aravinth Kumar, B. D. Sahoo, and A. Dutta. A wideband 2–5 GHz noise canceling subthreshold low noise amplifier. *IEEE Transactions on Circuits and Systems II*, 65(7):834–838, 2018.

[80] B. Guo, J. Chen, L. Li, H. Jin, and G. Yang. A wideband noise-canceling CMOS LNA with enhanced linearity by using complementary nMOS and pMOS configurations. *IEEE Journal of Solid-State Circuits*, 52(5):1331–1344, 2017.

[81] M. Parvizi, K. Allidina, and M. N. El-Gamal. Short channel output conductance enhancement through forward body biasing to realize a 0.5 v 250 μW 0.6–4.2 GHz current-reuse CMOS LNA. *IEEE Journal of Solid-State Circuits*, 51(3):574–586, 2016.

[82] M. Parvizi, K. Allidina, and M. N. El-Gamal. An ultra-low-power wideband inductorless CMOS LNA with tunable active shunt-feedback. *IEEE Transactions on Microwave Theory and Techniques*, 64(6):1843–1853, 2016.

[83] S. Bagga, A. L. Mansano, W. A. Serdijn, J. R. Long, K. van Hartingsveldt, and K. Philips. A frequency-selective broadband low-noise amplifier with double-loop transformer feedback. *IEEE Transactions on Circuits and Systems I: Regular Papers*, 61(6):1883–1891, 2014.

[84] J. Y.-C. Liu, J.-S. Chen, C. Hsia, P.-Y. Yin, and C.-W. Lu. A wideband inductorless single-to-differential LNA in 0.18 ⁻m CMOS technology for digital TV receivers. *IEEE Microwave and Wireless Components Letters*, 24(7):472–474, 2014.

[85] J. W. Park and B. Razavi. A harmonic-rejecting CMOS LNA for broadband radios. *IEEE Journal of Solid-State Circuits*, 48(4):1072–1084, 2013.

[86] K. H. Chen and S. I. Liu. Inductorless wideband CMOS low-noise amplifiers using noise cancelling technique. *IEEE Transactions on Circuits and Systems I*, 59(59):305–315, 2012.

[87] F. Belmas, F. Hameau, and J.-M. Fournier. A low power inductorless LNA with double G_m enhancement in 130 nm CMOS. *IEEE Journal of Solid-State Circuits*, 47(5):1094–1103, 2012.

[88] W. Zhuo, X. Li, S. Shekhar, S. H K Embabi, J. Pineda de Gyvez, D. J. Allstot, and E. Sanchez-Sinencio. A capacitor cross-coupled common-gate low-noise amplifier. *IEEE Transactions on Circuits and Systems II*, 52(12):875–879, 2005.

[89] F. Bruccoleri, E.A.M. Klumperink, and B. Nauta. Generating all two-MOS-transistor amplifiers leads to new wide-band LNAs. *IEEE Journal of Solid-State Circuits*, 36(7):1032–1040, 2001.

[90] F. Bruccoleri, E. A. M. Klumperink, and B. Nauta. Generating all two-MOS-transistor amplifiers leads to new wide-band LNAs. *IEEE Journal of Solid-State Circuits*, 36(7):1032–1040, 2001.

[91] A. Geis, J. Ryckaert, L. Bos, G. Vandersteen, Y. Rolain, and J. Craninckx. A 0.5 mm² power-scalable 0.5–3.8-GHz CMOS DT-SDR receiver with second-order RF band-pass sampler. *IEEE Journal of Solid-State Circuits*, 45(11):2375–2387, 2010.

[92] M. Tohidian, I. Madadi, and R. B. Staszewski. A fully integrated highly reconfigurable discrete-time superheterodyne receiver. In *Proceedings of the 2014 IEEE International Solid-State Circuits Conference – Digest of Technical Papers*, pages 1–3, 2014.

[93] A. Geis, J. Ryckaert, J. Borremans, G. Vandersteen, Y. Rolain, and J. Craninckx. A compact low power SDR receiver with 0.5–20MHz baseband sampled filter. In

Proceedings of the 2009 IEEE Radio Frequency Integrated Circuits Symposium, pages 285–288, 2009.

[94] S. Kiriaki, T. L. Viswanathan, G. Feygin, et al. A 160-MHz analog equalizer for magnetic disk read channels. *IEEE Journal of Solid-State Circuits*, 32(11):1839–1850, 1997.

[95] G. T. Uehara and P. R. Gray. A 100 MHz A/D interface for PRML magnetic disk read channels. *IEEE Journal of Solid-State Circuits*, 29(12):1606–1613, 1994.

[96] S.-S. Lee and C. A. Laber. A BiCMOS continuous-time filter for video signal processing applications. *IEEE Journal of Solid-State Circuits*, 33(9):1373–1382, 1998.

[97] F. Yang and C. C. Enz. A low-distortion BiCMOS seventh-order Bessel filter operating at 2.5 V supply. *IEEE Journal of Solid-State Circuits*, 31(3):321–330, 1996.

[98] S. D'Amico, M. Conta, and A. Baschirotto. A 4.1-mW 10-MHz fourth-order source-follower-based continuous-time filter with 79-dB DR. *IEEE Journal of Solid-State Circuits*, 41(12):2713–2719, 2006.

[99] T. Y. Lo, C. C. Hung, and M. Ismail. A wide tuning range Grm m -C filter for multimode CMOS direct-conversion wireless receivers. *IEEE Journal of Solid-State Circuits*, 44(9):2515–2524, 2009.

[100] A. Pirola, A. Liscidini, and R. Castello. Current-mode, WCDMA channel filter with in-band noise shaping. *IEEE Journal of Solid-State Circuits*, 45(9):1770–1780, 2010.

[101] M. S. Savadi Oskooei, N. Masoumi, M. Kamarei, and H. Sjoland. A CMOS 4.35-mW +22-dBm IIP3 continuously tunable channel select filter for WLAN/WiMAX Receivers. *IEEE Journal of Solid-State Circuits*, 46(6):1382–1391, 2011.

[102] A. Vasilopoulos, G. Vitzilaios, G. Theodoratos, and Y. Papananos. A low-power wide-band reconfigurable integrated active-RC filter with 73 dB SFDR. *IEEE Journal of Solid-State Circuits*, 41(9):1997–2008, 2006.

[103] S. Kousai, M. Hamada, R. Ito, and T. Itakura. A 19.7 MHz, fifth-order active-RC Chebyshev LPF for draft IEEE802.11n with automatic quality-factor tuning scheme. *IEEE Journal of Solid-State Circuits*, 42(11):2326–2337, 2007.

[104] M. Tohidian, I. Madadi, and R. B. Staszewski. A 2mw 800ms/s 7th-order discrete-time IIR filter with 400khz-to-30mhz BW and 100db stop-band rejection in 65nm CMOS. In *Proceedings of the 2013 IEEE International Solid-State Circuits Conference – Digest of Technical Papers*, pages 174–175, 2013.

[105] M. Ghaderi, J. Nossek, and G. Temes. Narrow-band switched-capacitor bandpass filters. *IEEE Transactions on Circuits and Systems*, 29(8):557–572, 1982.

[106] R. Winoto, G. Hueber B. Nikolic, and R. B. Staszewski. "Discrete time processing of RF signals." In *Multi-Mode/Multi-Band RF Transceivers for Wireless Communications: Advanced Techniques, Architectures, and Trends. Wiley-IEEE Press*, 29:219–245, 2011.

[107] J. Yuan. A charge sampling mixer with embedded filter function for wireless applications. In *ICMMT 2000. 2000 2nd International Conference on Microwave and Millimeter Wave Technology Proceedings (Cat. No.00EX364)*, pages 315–318, 2000.

[108] Y. S. Poberezhskiy and G. Y. Poberezhskiy. Corrections to "Sampling and Signal Reconstruction Circuits Performing Internal Antialiasing Filtering and Their Influence on the Design of Digital Receivers and Transmitters." *IEEE Transactions on Circuits and Systems I: Regular Papers*, 51(6):1234–1234, 2004.

[109] R. Gregorian and G. C. Temes. *Analog MOS Integrated Circuits for Signal Processing*. Wiley, New York, 1986.

[110] B. Nauta. A CMOS transconductance-C filter technique for very high frequencies. *IEEE Journal of Solid-State Circuits*, 27(2):142–153, 1992.

[111] A. Mirzaei, X. Chen, A. Yazdi, J. Chiu, J. Leete, and H. Darabi. A frequency translation technique for SAW-Less 3G receivers. In *2009 Symposium on VLSI Circuits*, pages 280–281, 2009.

[112] A. Geis. Discrete-time receiver topologies for SDR. PhD thesis, 2010.

[113] M. Kitsunezuka, T. Tokairin, T. Maeda, and M. Fukaishi. A low-IF/Zero-IF reconfigurable analog baseband IC with an I/Q imbalance cancellation scheme. *IEEE Journal of Solid-State Circuits*, 46(3):572–582, 2011.

[114] M. Tohidian, I. Madadi, and R.B. Staszewski. A fully integrated highly reconfigurable discrete-time superheterodyne receiver. In *Proceedings of the 2014 IEEE International Solid-State Circuits Conference – Digest of Technical Papers*, pages 72–73, 2014.

[115] L. E. Franks and I. W. Sandberg. An alternative approach to the realization of network transfer functions: The N-Path filter. *Bell Labs Technical Journal*, 39(5):1321–1350, 1960.

[116] A. Mirzaei, H. Darabi, and D. Murphy. Architectural evolution of integrated M-phase high-Q bandpass filters. *IEEE Transactions on Circuits and Systems I*, 59(1):52–65, 2012.

[117] A. Ghaffari, E. A. M. Klumperink, M. C. M. Soer, and B. Nauta. Tunable High-Q N-path band-pass filters: Modeling and verification. *IEEE Journal of Solid-State Circuits*, 46(5):998–1010, May 2011.

[118] M. Darvishi, R. van der Zee, and B. Nauta. Design of active N-path filters. *IEEE Journal of Solid-State Circuits*, 48(12):2962–2976, 2013.

[119] M. Darvishi, R. van der Zee, E. A. M. Klumperink, and B. Nauta. Widely tunable 4th order switched g_m-C band-pass filter based on n-path filters. *IEEE Journal of Solid-State Circuits*, 47(12):3105–3119, 2012.

[120] A. Mirzaei and H. Darabi. Analysis of imperfections on performance of 4-phase passive-mixer-based high-Q bandpass filters in SAW-Less receivers. *IEEE Transactions on Circuits and Systems*, 58(5):879–892, May 2011.

[121] G. Hueber and R. B. Staszewski. "Discrete-time processing of RF signals" in *Multi-Mode/Multi-Band RF Transceivers for Wireless Communications: Advanced Techniques, Architectures, and Trends*, pages 219–245. John Wiley & Sons, Inc., 2011.

[122] S. Karvonen, T. A. D. Riley, S. Kurtti, and J. Kostamovaara. A quadrature charge-domain sampler with embedded FIR and IIR filtering functions. *IEEE Journal of Solid-State Circuits*, 41(2):507–515, 2006.

[123] R. B. Staszewski, J. Wallberg, S. Rezeq, et al. All-digital PLL and GSM/EDGE transmitter in 90nm CMOS. In *Proceedings of the IEEE International Solid-State Circuits Conference*, pages 316–317, 2005.

[124] K. O. Estimation methods for quality factors of inductors fabricated in silicon integrated circuit process technologies. *IEEE Journal of Solid-State Circuits*, pages 1249–1252, 1998.

[125] A. Mirzaei, H. Darabi, A. Yazdi, Z. Zhou, E. Chang, and P. Suri. A 65 nm CMOS quad-band SAW-Less receiver SoC for GSM/GPRS/EDGE. *IEEE Journal of Solid-State Circuits*, 46(4):950–964, 2011.

[126] H. Darabi. A blocker filtering technique for SAW-Less wireless receivers. *IEEE Journal of Solid-State Circuits*, 42(12):2766–2773, 2007.

[127] A. Ghaffari, E. Klumperink, and B. Nauta. 8-path tunable RF notch filters for blocker suppression. In *Proceedings of the 2012 IEEE International Solid-State Circuits Conference – Digest of Technical Papers*, pages 76 –78, 2012.

[128] I. Madadi, M. Tohidian, and R. B. Staszewski. Analysis and design of I/Q charge-sharing band-pass-filter for superheterodyne receivers. *IEEE Transactions on Circuits and Systems I: Regular Papers*, 62(8):2114–2121, 2015.

[129] E. Morifuji, H. S. Momose, T. Ohguro, et al. Future perspective and scaling down roadmap for RF CMOS. *1999 IEEE Symposium on VLSI Technology and Circuits – Digest of Technical Papers (IEEE Cat. No.99CH36325)*, pages 163–164, 1999.

[130] K. Lee, I. Nam, I. Kwon, J. Gil, K. Han, S. Park, and B. I. Seo. The impact of semiconductor technology scaling on CMOS RF and digital circuits for wireless application. *IEEE Transactions on Electron Devices*, 52(7):1415–1422, 2005.

[131] L. F. Tiemeijer, R. J. Havens, R. De Kort, et al. Record RF performance of standard 90 nm CMOS technology. *IEEE International Electron Devices Meeting 2004 – IEDM Technical Digest*, pages 441–444, 2004.

[132] P. H. Woerlee, M. J. Knitel, R.van Langevelde, et al. RF-CMOS performance trends. *IEEE Transactions on Electron Devices*, 48:1776–1782, 2001.

[133] B. Verbruggen, M. Iriguchi, M. de la Guia Solaz, et al. A 2.1 mW 11b 410 MS/s dynamic pipelined SAR ADC with background calibration in 28nm digital CMOS. In *2013 IEEE Symposium on VLSI Circuits*, pages C268–C269, 2013.

[134] G. Hueber and R. B. Staszewski. *Multi-Mode/Multi-Band RF Transceivers for Wireless Communications: Advanced Techniques, Architectures, and Trends*. Wiley, 2011.

[135] J. A. Weldon, J. C. Rudell, R. S. Narayanaswami, M. Otsuka, S. Dedieu, and P. R. Gray. A 1.75 GHz highly-integrated narrow-band CMOS transmitter with harmonic-rejection mixers. In *Proceedings of the 2001 IEEE International Solid-State Circuits Conference – Digest of Technical Papers ISSCC (Cat. No.01CH37177)*, volume 36, pages 160–161, 2003.

[136] T. Forbes, W. G. Ho, and R. Gharpurey. Design and analysis of harmonic rejection mixers with programmable LO frequency. *IEEE Journal of Solid-State Circuits*, 48(10):2363–2374, 2013.

[137] H. Darabi. Highly integrated and tunable RF front-ends for reconfigurable multi-band transceivers. *Cust. Integr. Circuits Conf. (CICC), 2010 IEEE*, 2010.

[138] D. Murphy, H. Darabi, and H. Xu. A noise-cancelling receiver with enhanced resilience to harmonic blockers. In *Proceedings of the IEEE International Solid-State Circuits Conference – Digest of Technical Papers*, volume 57, pages 68–69, 2014.

Index

Printed in the United States
by Baker & Taylor Publisher Services